Using Time Series to Analyze Long-Range Fractal Patterns

Quantitative Applications in the Social Sciences

A SAGE PUBLICATIONS SERIES

1. Analysis of Variance, 2nd Edition *Iversen/Norpoth*
2. Operations Research Methods *Nagel/Neef*
3. Causal Modeling, 2nd Edition *Asher*
4. Tests of Significance *Henkel*
5. Cohort Analysis, 2nd Edition *Glenn*
6. Canonical Analysis and Factor Comparison *Levine*
7. Analysis of Nominal Data, 2nd Edition *Reynolds*
8. Analysis of Ordinal Data *Hildebrand/Laing/Rosenthal*
9. Time Series Analysis, 2nd Edition *Ostrom*
10. Ecological Inference *Langbein/Lichtman*
11. Multidimensional Scaling *Kruskal/Wish*
12. Analysis of Covariance *Wildt/Ahtola*
13. Introduction to Factor Analysis *Kim/Mueller*
14. Factor Analysis *Kim/Mueller*
15. Multiple Indicators *Sullivan/Feldman*
16. Exploratory Data Analysis *Hartwig/Dearing*
17. Reliability and Validity Assessment *Carmines/Zeller*
18. Analyzing Panel Data *Markus*
19. Discriminant Analysis *Klecka*
20. Log-Linear Models *Knoke/Burke*
21. Interrupted Time Series Analysis *McDowall/McCleary/Meidinger/Hay*
22. Applied Regression, 2nd Edition *Lewis-Beck/Lewis-Beck*
23. Research Designs *Spector*
24. Unidimensional Scaling *McIver/Carmines*
25. Magnitude Scaling *Lodge*
26. Multiattribute Evaluation *Edwards/Newman*
27. Dynamic Modeling *Huckfeldt/Kohfeld/Likens*
28. Network Analysis *Knoke/Kuklinski*
29. Interpreting and Using Regression *Achen*
30. Test Item Bias *Osterlind*
31. Mobility Tables *Hout*
32. Measures of Association *Liebetrau*
33. Confirmatory Factor Analysis *Long*
34. Covariance Structure Models *Long*
35. Introduction to Survey Sampling, 2nd Edition *Kalton*
36. Achievement Testing *Bejar*
37. Nonrecursive Causal Models *Berry*
38. Matrix Algebra *Namboodiri*
39. Introduction to Applied Demography *Rives/Serow*
40. Microcomputer Methods for Social Scientists, 2nd Edition *Schrodt*
41. Game Theory *Zagare*
42. Using Published Data *Jacob*
43. Bayesian Statistical Inference *Iversen*
44. Cluster Analysis *Aldenderfer/Blashfield*
45. Linear Probability, Logit, and Probit Models *Aldrich/Nelson*
46. Event History and Survival Analysis, 2nd Edition *Allison*
47. Canonical Correlation Analysis *Thompson*
48. Models for Innovation Diffusion *Mahajan/Peterson*
49. Basic Content Analysis, 2nd Edition *Weber*
50. Multiple Regression in Practice *Berry/Feldman*
51. Stochastic Parameter Regression Models *Newbold/Bos*
52. Using Microcomputers in Research *Madron/Tate/Brookshire*
53. Secondary Analysis of Survey Data *Kiecolt/Nathan*
54. Multivariate Analysis of Variance *Bray/Maxwell*
55. The Logic of Causal Order *Davis*
56. Introduction to Linear Goal Programming *Ignizio*
57. Understanding Regression Analysis, 2nd Edition *Schroeder/Sjoquist/Stephan*
58. Randomized Response and Related Methods, 2nd Edition *Fox/Tracy*
59. Meta-Analysis *Wolf*
60. Linear Programming *Feiring*
61. Multiple Comparisons *Klockars/Sax*
62. Information Theory *Krippendorff*
63. Survey Questions *Converse/Presser*
64. Latent Class Analysis *McCutcheon*
65. Three-Way Scaling and Clustering *Arabie/Carroll/DeSarbo*
66. Q Methodology, 2nd Edition *McKeown/Thomas*
67. Analyzing Decision Making *Louviere*
68. Rasch Models for Measurement *Andrich*
69. Principal Components Analysis *Dunteman*
70. Pooled Time Series Analysis *Sayrs*
71. Analyzing Complex Survey Data, 2nd Edition *Lee/Forthofer*
72. Interaction Effects in Multiple Regression, 2nd Edition *Jaccard/Turrisi*
73. Understanding Significance Testing *Mohr*
74. Experimental Design and Analysis *Brown/Melamed*
75. Metric Scaling *Weller/Romney*
76. Longitudinal Research, 2nd Edition *Menard*
77. Expert Systems *Benfer/Brent/Furbee*
78. Data Theory and Dimensional Analysis *Jacoby*
79. Regression Diagnostics, 2nd Edition *Fox*
80. Computer-Assisted Interviewing *Saris*
81. Contextual Analysis *Iversen*
82. Summated Rating Scale Construction *Spector*
83. Central Tendency and Variability *Weisberg*
84. ANOVA: Repeated Measures *Girden*
85. Processing Data *Bourque/Clark*
86. Logit Modeling *DeMaris*
87. Analytic Mapping and Geographic Databases *Garson/Biggs*
88. Working With Archival Data *Elder/Pavalko/Clipp*
89. Multiple Comparison Procedures *Toothaker*
90. Nonparametric Statistics *Gibbons*
91. Nonparametric Measures of Association *Gibbons*
92. Understanding Regression Assumptions *Berry*
93. Regression With Dummy Variables *Hardy*
94. Loglinear Models With Latent Variables *Hagenaars*
95. Bootstrapping *Mooney/Duval*
96. Maximum Likelihood Estimation *Eliason*
97. Ordinal Log-Linear Models *Ishii-Kuntz*
98. Random Factors in ANOVA *Jackson/Brashers*
99. Univariate Tests for Time Series Models *Cromwell/Labys/Terraza*

Quantitative Applications in the Social Sciences

A SAGE PUBLICATIONS SERIES

100. Multivariate Tests for Time Series Models *Cromwell/Hannan/Labys/Terraza*
101. Interpreting Probability Models: Logit, Probit, and Other Generalized Linear Models *Liao*
102. Typologies and Taxonomies *Bailey*
103. Data Analysis: An Introduction *Lewis-Beck*
104. Multiple Attribute Decision Making *Yoon/Hwang*
105. Causal Analysis With Panel Data *Finkel*
106. Applied Logistic Regression Analysis, 2nd Edition *Menard*
107. Chaos and Catastrophe Theories *Brown*
108. Basic Math for Social Scientists: Concepts *Hagle*
109. Basic Math for Social Scientists: Problems and Solutions *Hagle*
110. Calculus *Iversen*
111. Regression Models: Censored, Sample Selected, or Truncated Data *Breen*
112. Tree Models of Similarity and Association *Corter*
113. Computational Modeling *Taber/Timpone*
114. LISREL Approaches to Interaction Effects in Multiple Regression *Jaccard/Wan*
115. Analyzing Repeated Surveys *Firebaugh*
116. Monte Carlo Simulation *Mooney*
117. Statistical Graphics for Univariate and Bivariate Data *Jacoby*
118. Interaction Effects in Factorial Analysis of Variance *Jaccard*
119. Odds Ratios in the Analysis of Contingency Tables *Rudas*
120. Statistical Graphics for Visualizing Multivariate Data *Jacoby*
121. Applied Correspondence Analysis *Clausen*
122. Game Theory Topics *Fink/Gates/Humes*
123. Social Choice: Theory and Research *Johnson*
124. Neural Networks *Abdi/Valentin/Edelman*
125. Relating Statistics and Experimental Design: An Introduction *Levin*
126. Latent Class Scaling Analysis *Dayton*
127. Sorting Data: Collection and Analysis *Coxon*
128. Analyzing Documentary Accounts *Hodson*
129. Effect Size for ANOVA Designs *Cortina/Nouri*
130. Nonparametric Simple Regression: Smoothing Scatterplots *Fox*
131. Multiple and Generalized Nonparametric Regression *Fox*
132. Logistic Regression: A Primer *Pampel*, 2nd Edition
133. Translating Questionnaires and Other Research Instruments: Problems and Solutions *Behling/Law*
134. Generalized Linear Models: A Unified Approach, 2nd Edition *Gill/Torres*
135. Interaction Effects in Logistic Regression *Jaccard*
136. Missing Data *Allison*
137. Spline Regression Models *Marsh/Cormier*
138. Logit and Probit: Ordered and Multinomial Models *Borooah*
139. Correlation: Parametric and Nonparametric Measures *Chen/Popovich*
140. Confidence Intervals *Smithson*
141. Internet Data Collection *Best/Krueger*
142. Probability Theory *Rudas*
143. Multilevel Modeling, 2nd Edition *Luke*
144. Polytomous Item Response Theory Models *Ostini/Nering*
145. An Introduction to Generalized Linear Models *Dunteman/Ho*
146. Logistic Regression Models for Ordinal Response Variables *O'Connell*
147. Fuzzy Set Theory: Applications in the Social Sciences *Smithson/Verkuilen*
148. Multiple Time Series Models *Brandt/Williams*
149. Quantile Regression *Hao/Naiman*
150. Differential Equations: A Modeling Approach *Brown*
151. Graph Algebra: Mathematical Modeling With a Systems Approach *Brown*
152. Modern Methods for Robust Regression *Andersen*
153. Agent-Based Models, 2nd Edition *Gilbert*
154. Social Network Analysis, 3rd Edition *Knoke/Yang*
155. Spatial Regression Models, 2nd Edition *Ward/Gleditsch*
156. Mediation Analysis *Iacobucci*
157. Latent Growth Curve Modeling *Preacher/Wichman/MacCallum/Briggs*
158. Introduction to the Comparative Method With Boolean Algebra *Caramani*
159. A Mathematical Primer for Social Statistics *Fox*, 2nd Edition
160. Fixed Effects Regression Models *Allison*
161. Differential Item Functioning, 2nd Edition *Osterlind/Everson*
162. Quantitative Narrative Analysis *Franzosi*
163. Multiple Correspondence Analysis *LeRoux/Rouanet*
164. Association Models *Wong*
165. Fractal Analysis *Brown/Liebovitch*
166. Assessing Inequality *Hao/Naiman*
167. Graphical Models and the Multigraph Representation for Categorical Data *Khamis*
168. Nonrecursive Models *Paxton/Hipp/Marquart-Pyatt*
169. Ordinal Item Response Theory *Van Schuur*
170. Multivariate General Linear Models *Haase*
171. Methods of Randomization in Experimental Design *Alferes*
172. Heteroskedasticity in Regression *Kaufman*
173. An Introduction to Exponential Random Graph Modeling *Harris*
174. Introduction to Time Series Analysis *Pickup*
175. Factorial Survey Experiments *Auspurg/Hinz*
176. Introduction to Power Analysis: Two-Group Studies *Hedberg*
177. Linear Regression: A Mathematical Introduction *Gujarati*
178. Propensity Score Methods and Applications *Bai/Clark*
179. Multilevel Structural Equation Modeling *Silva/Bosancianu/Littvay*
180. Gathering Social Network Data *adams*
181. Generalized Linear Models for Bounded and Limited Quantitative Variables, *Smithson and Shou*
182. Exploratory Factor Analysis, *Finch*
183. Multidimensional Item Response Theory, *Bonifay*
184. Argument-Based Validation in Testing and Assessment, *Chapelle*
185. Using Time Series to Analyze Long-Range Fractal Patterns, *Koopmans*
186. Understanding Correlation Matrices *Hadd/Rodgers*
187. Rasch Models for Solving Measurement Problems *Engelhard Jr./Wang*

Using Time Series to Analyze Long-Range Fractal Patterns

Matthijs Koopmans
Mercy College

Los Angeles | London | New Delhi
Singapore | Washington DC | Melbourne

For information:

SAGE Publications, Inc.
2455 Teller Road
Thousand Oaks, California 91320
E-mail: order@sagepub.com

SAGE Publications Ltd.
1 Oliver's Yard
55 City Road
London EC1Y 1SP
United Kingdom

SAGE Publications India Pvt. Ltd.
B 1/I 1 Mohan Cooperative Industrial Area
Mathura Road, New Delhi 110 044
India

SAGE Publications Asia-Pacific Pte. Ltd.
18 Cross Street #10-10/11/12
China Square Central
Singapore 048423

Library of Congress Cataloging-in-Publication Data

Names: Koopmans, Matthijs, author.

Title: Using time series to analyze long range fractal patterns / Matthijs Koopmans, Mercy College.

Description: Los Angeles : SAGE, [2021] | Includes bibliographical references.

Identifiers: LCCN 2020031198 | ISBN 9781544361420 (paperback) | ISBN 9781544361437 (epub) | ISBN 9781544361444 (epub) | ISBN 9781544361413 (ebook)

Subjects: LCSH: Social sciences—Statistical methods. | Time-series analysis. | Fractal analysis.

Classification: LCC HA30.3 .K66 2021 | DDC 519.5/5—dc23

LC record available at https://lccn.loc.gov/2020031198

This book is printed on acid-free paper.

Acquisitions Editor: Helen Salmon
Editorial Assistant: Elizabeth Cruz
Production Editor: Natasha Tiwari
Copy Editor: QuADS Prepress Pvt. Ltd.
Typesetter: Hurix Digital
Proofreader: Jeff Bryant
Indexer: Integra
Cover Designer: Candice Harman
Marketing Manager: Victoria Velasquez

20 21 22 23 24 10 9 8 7 6 5 4 3 2 1

CONTENTS

Series Editor Introduction   ix

Acknowledgments   xi

About the Author   xii

Chapter 1: Introduction   1
  A. Limitations of Traditional Approaches   4
  B. Long-Range Dependencies   7
  C. The Search for Complexity   9
  D. Plan of the Book   12

**Chapter 2: Autoregressive Fractionally Integrated Moving
Average or Fractional Differencing**   15
  A. Basic Results in Time Series Analysis   15
    Seasonal Patterns   18
    Integration   19
    Testing for Stationarity   21
  B. Long-Range Dependencies   22
  C. Application of the Models to Real Data   24
    Competitive Modeling Strategies   30
    Differencing Detrended Data   37
    Analyzing the Residuals   38
    Interpretation of the Differencing Parameter 39
    The Hurst Exponent   40
  D. Chapter Summary and Reflection   41

Chapter 3: Power Spectral Density Analysis   43
  A. From the Time Domain to the Frequency Domain   44
    Amplitudes and Relative Frequencies 44
    The Fourier Transform   46
    Periodograms   49
    Power Spectral Density   50
  B. Spectral Density in Real Data   52

C. Fractional Estimates of Gaussian Noise and
    Brownian Motion   55
D. Chapter Summary and Reflection   57

Chapter 4: Related Methods in the Time and
Frequency Domains   **59**
  A. Estimating Fractal Variance   59
    Detrended Fluctuation Analysis   60
    Rescaled Range Analysis   63
    Higuchi's Fractal Dimension   65
    Related Approaches   69
  B. Spectral Regression   69
  C. The Hurst Exponent Revisited   73
  D. Chapter Summary and Reflection   74

Chapter 5: Variations on the Fractality Theme   **77**
  A. Sensitive Dependence on Initial Conditions   78
  B. The Multivariate Case   78
  C. Regular Long-Range Processes and Nested Regularity   80
  D. The Impact of Interventions   81

Chapter 6: Conclusion   **85**
  A. Benefits and Drawbacks of Fractal Analysis   86
  B. Interpretation of Parameters in Terms
    of Complexity Theory   89
  C. A Note About the Software and Its Use   90

References   **93**

Appendix   **101**

Index   **103**

SERIES EDITOR INTRODUCTION

It is with great pleasure that I introduce *Using Time Series to Analyze Long-Range Fractal Patterns* by Matthijs Koopmans. This book presents methods for describing and analyzing dependency and irregularity in long time series of up to 800 observations or even more. Irregularity refers to cycles that are similar in appearance, but unlike seasonal patterns more familiar to social scientists, repeated over a time scale that is not fixed—that is, fractal patterns. Until now, the application of these methods has mainly involved analysis of dynamical systems outside of the social sciences—for example, in biology, hydrology, chemistry, or physics. This book makes it possible for social scientists to explore and document fractal patterns in dynamical social systems.

Using Time Series to Analyze Long-Range Fractal Patterns concentrates on two general approaches to irregularity in long time series: autoregressive fractionally integrated moving average (ARFIMA) models, also known as fractional differencing, and power spectral density analysis (PSDA). The former focuses on time (the behavior of individual data points and their correlation to each other), the latter on frequencies (the detection of cycles that repeat at various frequencies and their contribution to overall variability of the series). The two approaches have complementary strengths. For example, fractional differencing enables the user to evaluate the contribution of fractional and regular patterns to model fit; this is not possible with PSDA. A strength of PSDA is its ability to model scale invariance of nonrandom patterns over the long range, not possible with ARFIMA. Because of their complementary features, Professor Koopmans advises that analysts use both approaches. Indeed, some of the methods are still in development. Professor Koopmans is careful to point out the unresolved issues and complexities while at the same time providing a solid foundation in the current state of the field.

Two kinds of examples are key to this book. First are simulations: white noise, short-run dependency, pink noise (frequencies have varying probabilities across the series), random walk, and seasonality. The simulations illustrate the patterns that might be encountered and serve as a benchmark for interpreting patterns in real data. Second are social science examples.

These include monthly unemployment figures for the United States from 1947 to 2017, daily attendance rates in an urban high school from 2010 to 2014, daily number of births to teens in Texas from 1964 to 1990, and weekly survey data on political orientation from 1978 to 1996 in the Netherlands. It is difficult to overstate the importance of social science examples to making a new method accessible, as all of us who learned differential equations through the lens of physics and chemistry can attest. In keeping with the book's primary focus, all of the examples are univariate. Data and R script to replicate the analyses are available on the accompanying website for this book at **https://study.sagepub.com/researchmethods/qass/koopmans-using-time-series**.

Data requirements for the methods described in this book are stringent. As Professor Koopmans points out, the need for information over the long range of a time series presents logistical challenges. This is true, especially with respect to data sources that social scientists traditionally use. For instance, it will be more than another decade before 800 monthly observations are available from our longest running longitudinal social science survey, the Panel Study of Income Dynamics. That said, the field is in the midst of a major transition in data sources used to study social and behavioral phenomena. Administrative data increasingly cover long periods of time, including the monthly unemployment statistics used as an example in the book. Even more exciting is the temporal detail capturable in Internet searches, tweets, and spatial movements tracked by GPS-enabled smartphones. With these and other new sources of data, behavior can be measured in days, hours, minutes, and seconds, with the length of the series no longer a concern. As data requirements recede as a challenge, increasingly important will be the social imagination needed to describe and especially interpret fractal patterns in the data.

This book will help get social scientists get started. With Professor Koopmans as guide, readers familiar with calculus and basic statistics can take their first steps to master the tools.

Barbara Entwisle
Series Editor

ACKNOWLEDGMENTS

We are grateful for the reviewers who provided valuable feedback in the development of this book:

John G. Bretting, University of Texas at El Paso

Courtney Brown, Emory University

I.-Ming Chiu, Rutgers University Camden

Mustafa Demir, State University of New York at Plattsburgh

Henry Hyunsuk Kim, Wheaton College

Warren E. Lacefield, Academic Software, Inc.

Matthew Lebo, Stony Brook University

Clayton Webb, University of Kansas

For their help gaining access to the data sets used in this book, the author thanks the following colleagues:

Barbara Dworkowitz, New York City Department of Education

Patricia Hamilton, Texas Women's University

Janet Box-Steffensmeier, Ohio State University

ABOUT THE AUTHOR

Matthijs Koopmans, professor of educational leadership, joined the faculty at Mercy College in 2011. Previously, he worked for several educational research organizations, including the Strategic Education Research Partnership Institute, Academy for Educational Development, and Metis Associates. He has taught at several colleges in the greater New York metropolitan region (Hofstra University, York College/City University of New York, Adelphi University, and Yeshiva University). As an independent contractor, he has conducted evaluations for MGT of America, Institute for Student Achievement, National Urban Technology Center, and Newark Public Schools. In scholarship, his areas of focus are the application of complexity theory to education, cause-and-effect relationships, and the estimation of fractal patterns in time series data. This book pursues the latter interest, which stems from a belief that the effective application of principles of complexity theory in the social sciences should include attempts to extend conventional parametric models to the estimation of dynamical processes. He published his research in numerous peer-reviewed journals and continues to present his work at national and international scholarly conferences. He is a founding editor of the *International Journal of Complexity in Education* and serves on the editorial board of *Nonlinear Dynamics, Psychology, and Life Sciences*. He earned his doctorate in 1988 from the Harvard Graduate School of Education.

1

INTRODUCTION

Human behavior is an evolving phenomenon, and the description of its evolution is therefore a pertinent concern. Why do behaviors change or stay the same? How does behavioral transformation work? Under what conditions does change occur? While social science research may or may not explicitly address these issues, they linger in the background of any study as a potential story about the underlying dynamics of the behavior observed. To deal with such dynamics, we need to inquire about the details of the timing of events, and its impact on the behavior we are interested in, as well as investigate natural fluctuations in behavior that occur irrespective of any input from the environment. Measures of seasonal changes in temperature and precipitation, for example, reflect the earth's annual trajectory around the sun and the tilt of its axis relative to that trajectory, leading to different seasonal patterns on the northern and southern hemispheres. Our daily and hourly recordings of temperature reflect and have helped us understand these underlying dynamics.

There are methodological as well as substantive reasons why social scientists might be interested in the time dependency of behavior. To make causal attributions, we often presume that effects are preceded by their causes, implying a time spectrum on which measurements concerning the two types of events are ordered successively. There may also be more complex feedback relationships in which behavioral outcomes affect the input conditions in a relational cycle that repeats—perhaps very often—for example, in an interactional exchange between two people in which tensions escalate into an argument. On the methodological side, there is the question how to characterize a distribution of findings if the data points of interest are not independent, as is commonly assumed in inferential statistics. If observations are sequentially ordered, the assumption of independent observations cannot be taken for granted. This book addresses one aspect of this issue, which is the detection and analysis of complex time-dependent patterns, such as fractals, in a set of repeated observations. Fractals are shapes made of parts whose appearances are similar to that of the whole that contains them (Feder, 1988), and they do not necessarily characterize time series but can also describe geometrical shapes (e.g., Lauwerier, 1991).

1

There are many good introductory texts on the market about time series analysis (e.g., Cryer & Chan, 2008; Shumway & Stoffer, 2011), as well as several volumes in the SAGE *Quantitative Applications in the Social Sciences* series *(QASS)* that focus on the various aspects of this approach (Brandt & Williams, 2006; McDowall et al., 1980; Ostrom, 1990; Pickup, 2014; Sayrs, 1989). However, within the QASS series, there are no contributions dealing specifically with long-range dependencies and irregularity. While Brown and Liebovitch's (2010) *Fractal Analysis* addresses self-similarity, and their work stems from the same intellectual tradition as the methodologies discussed here, their focus is on the geometrical rather than the temporal properties of fractals. They generally do not utilize time series in that context except for a four-page section toward the end of the book, in which they admit that "it really merits a volume on its own" (p. 68). Hence the present book.

Time series analysis has been the established approach to the analysis of time dependencies (e.g., Box & Jenkins, 1970), in which the ordering of observations across the time scale serves as a predictor to the time-dependent outcomes. This time scale could be defined in terms of years, months, weeks, days, hours, or seconds, and a time series analysis can then be used to decide over which number of lags outcomes can be predicted based on previous observations. The equation below expresses this relationship between a given value and its immediately preceding values at the first and second lags. The preceding values serve here as predictors of a subsequent one:

$$Y_t = \varphi_1 Y_{t-1} + \varphi_2 Y_{t-2} + e_t. \tag{1.1}$$

As in a conventional regression equation, the contribution to the variability in Y_t can be separately established for preceding values at the first and second lags in this model. It is presumed that a successful modeling effort results in the errors e_t being independent with a random normal distribution and a mean of zero.

A central construct in any time series analysis is *autocorrelation*, which expresses the statistical association of given measurements on a time series with neighboring measurements of that same variable (see Chapter 2, Equation 2.1, for the computational details). While a codified methodology to handle short-term dependencies in a time series such as those described by Equation 1.1 has been in place for quite some time now, our ability to model the influence of autocorrelations between data points that are further apart on the time scale has been a more recent development (see, e.g., Granger & Joyeux, 1980). Traditional time series was designed to estimate autocorrelations between observations that are in relatively close proximity on the series. The primary concern of this book is with autocorrelations between points that are further apart because they suggest a more complex pattern of within-subject relationships. Note that long-range patterns, as discussed in this book, refer to autocorrelations being far apart in terms of the number of lags, not necessarily the length of the absolute time range. Thus, over a spectrum of 5 minutes, we can show a long-range pattern if measurements are repeated every second, but not on a scale where they repeat every minute. Later on in this chapter, I will return to this point.

One scientific discipline in which correlations over the long range of a time scale are a pertinent concern is hydrology, which has, among other things, kept records of river discharges whose measurement results may depend not only on recent precipitation but also on earlier rainfall (Feder, 1988), hence creating the need for a long-range perspective on the data. The analysis of long-range patterns in time series has been successfully applied elsewhere as well, in areas as diverse as irregular heartbeat (Peng et al., 1993), fine motor coordination (Chen et al., 1997), the perception of reversible figures (Aks & Sprott, 2003), response times in cognitive task performance (Van Orden et al., 2005), self-esteem (Wong et al., 2014), births to teens in the state of Texas (Hamilton et al., 1997), high school attendance rates in New York City (Koopmans, 2018b), and the U.S. gross national product (Sowell, 1992). In addition, Mandelbrot (1997) discusses the relevance of fractal patterns in financial time series data (e.g., fluctuating cotton prices). In each of these cases, modeling self-similarity in the data adds predictability over and above the patterns of regularity handled by traditional time series analysis.

With respect to long-range dependencies in time series data, two scenarios can be distinguished. For one, it may be that correlations exist between observations that are far apart in a pattern that is highly regular—for example, yearly cycles in daily recordings or weekly cycles within annual ones. It is also possible that a time series contains an ongoing pattern of irregular cycles that repeat in an unpredictable manner. For the use of time series analysis to handle regular long-range processes, a lucid overview is provided by Hyndman and Athanasopoulos (2012). The irregular processes are the main concern of the present book. In the remainder of this chapter, I will discuss the preliminaries to such an analysis by describing the challenges that time-dependent data pose to traditional approaches to statistical inference irrespective of whether those dependencies are over the short or the long range of a time series. I will introduce the long-range perspective on time-dependent data and provide some context from the vantage point of complexity theory, which takes an active interest in the detection and interpretation of fractal patterns, and what they might tell us about the stability and change in the behavior of social and other systems.

A. Limitations of Traditional Approaches

Our ability to make statistical inferences is typically informed by cross-sectional data structures, where a sampling of independent observations is taken to represent the properties of a population of interest. Sample measures are then used to estimate population parameters. To ensure representativeness, each individual from the population needs to have an equal likelihood of being included in the sample that is drawn. This implies that observations of those individuals are independent of one another (Beran, 1994). There is little reason to expect that this presumption carries over to a sampling of within-subject measurements sampled from a larger time spectrum, such that data from a small number of time points characterizes behavior over the entire spectrum. The question of dependency between such observations therefore goes to the heart of our ability to draw reliable conclusions from the data we collect (Molenaar, 2004), and the absence of any dependency among observations within subjects needs to be separately established in such cases. Time series analysis is expressly designed to do so.

Analyzing the dependency between observations over time is not only a matter of statistical prudence (Clarke & Lebo, 2003), but it can be of intrinsic interest as well. Our daily newspapers, for example, report weather patterns (temperature, precipitation) and mortgage and interest rates as lengthy time series on a regular basis. It is not difficult to appreciate that when traditional summary statistics such as means, ranges, histograms, and regression lines are generated, the summary measures fail to capture what is interesting about these data—for example, seasonal changes in the weather and temperature shifts within those seasons. Time series analysis models the effect of time by evaluating whether the variability in a data trajectory is random or is somehow dependent on the ordering of the data points. To appreciate why this matters, it is instructive to take a look at an example of nonrandomness in time series data shown in Figure 1.1. These data were taken from a study by McKuen et al. (1989), using data that were later expanded on and reanalyzed by Box-Steffensmeier and Smith (1998). The concern of these two studies is the degree of stability of political phenomena in the United States, and constituents' shifting attitudes toward the political system over time, exemplified here with data from U.S. presidential terms from 1953 through 1993 (i.e., from Eisenhower through George H. W. Bush). Based on findings from quarterly administrations of the Gallup poll to a random sample of respondents, Figure 1.1 shows the fluctuation in three types of outcomes pertaining to this question: (1) presidential approval, (2) consumer confidence, and (3) macropartisanship ratings. Macropartisanship is defined here as the aggregate percentage of respondents out of all responding party identifiers who identify as Democrats.

The plots in Figure 1.1 show considerable volatility in the distribution across the time scale in all three instances. It can also be seen that these measures do not fluctuate in a random fashion but, instead, show distinct

Figure 1.1 Quarterly Survey Ratings of Presidential Approval, Consumer Sentiment and Macro-Partisanship from 1953 through 1993

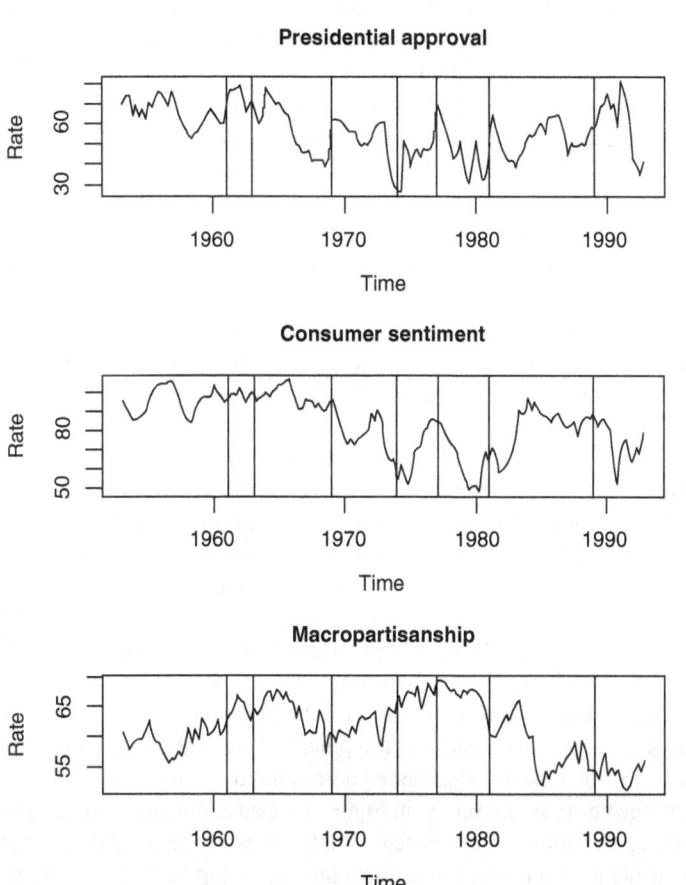

Source. Adapted from McKuen et al. (1989).

Note. Presidential periods are distinguished by vertical lines as follows: 1953–1961 Dwight D. Eisenhower (R); 1961–1963 John F. Kennedy (D), 1963–1969 Lyndon B. Johnson (D), 1969–1974 Richard M. Nixon (R), 1974–1977 Gerald R. Ford (R), 1977–1981 Jimmy E. Carter (D), 1981–1989 Ronald W. Reagan (R), 1989–1993 George H. W. Bush (R). R = Republican; D = Democrat.

patterns for different presidential periods, illustrating the importance of the time at which these measurements were taken. For example, the drop in the approval ratings for the Johnson administration in the late 1960s is not accompanied by a similar drop in consumer confidence during that same period. It can also be seen in the figure that during the Nixon and Ford administrations, there is a continuing upward trend in macropartisanship toward a democratic leaning, which starts to decline during the Carter years and falls dramatically during the Reagan years. The approval ratings of this latter president fluctuate in the course of his term but do not show the kind of downward slope observed during the Johnson, Nixon, Carter, and George H. W. Bush administrations. The timing of the changes in these three measures are suggestive of how the electorate responds to circumstances affecting the political system, such as perhaps the Vietnam War (Johnson), Watergate (Nixon), the Iran hostage crisis (Carter), or the broken "read my lips" tax pledge (George H. W. Bush). The plots in Figure 1.1 also suggest that consumer confidence and presidential approval, but not macropartisanship, covary during the Reagan, Nixon, Ford, and Carter years, but not so much during the terms of the other presidents. Furthermore, macropartisanship covaries with presidential approval during the Eisenhower, Kennedy, and Johnson administrations, but not beyond.

These trends are very interesting, but we would not have been able to uncover them if the ordering of the data points over time had not been considered. Computing average approval ratings across the entirety of the terms of these presidents would not describe their differences very well. Do approval ratings go up or down over the course of a single presidential term? Is a downward slope during one presidential term followed by a recovery at the onset of the next presidency? Are there predictable cycles within presidential terms for these political indicators? Is there an interdependency between approval ratings, on the one hand, and consumer confidence and macropartisanship on the other? None of these questions can be answered if the data are aggregated across the time spectrum.

Data such as those presented in Figure 1.1 also allow one to investigate the underlying dynamics of the system of interest and evaluate the applicability of the general principles of dynamical theories—that is, theories that explicitly concern themselves with processes of stability and transformation in systems (e.g., Kauffman, 1993; Prigogine & Stengers, 1984; von Berthalanffy, 1968). Box-Steffensmeier and Smith (1998) frame their reanalysis of these data in terms of stability and change in the political system in the United States and the extent to which indicators of political preference display volatility. These authors also examine the extent to which the three trajectories are linked such that fluctuations in macropartisanship may be linked to variations in presidential approval and consumer sentiment. In this way, the dynamical interplay between related processes can be evaluated to enhance our understanding of the political system and its workings.

B. Long-Range Dependencies

Time series analysis has been mostly concerned with the analysis of dependencies between observations over the short range, which is to say between observations that are located near each other on the time scale, such that we may estimate, for example, whether days of the week (7 lags) or months of the year (12 lags) are correlated. Accordingly, we typically collect data in time chunks that are relatively small, say over a few weeks or a few months. However, there are instances in which the dependency between data points plays out over a much longer time period, and such a dynamic may not be discernible if a smaller sampling of time points is considered. Such would be the case, for instance, in the study of natural phenomena, such as the length of tree rings or the volume of water discharged into the sea by rivers (Beran, 1994)—processes that play out over much longer periods.

The relevance of using a wider time frame for the study of long-range phenomena is illustrated in Figure 1.2, which shows the daily reported births to teens in the state of Texas, part of a data set maintained by the Texas Department of Health (see Hamilton et al., 1997). The top panel shows the recordings of such births over a 6-month period (the first half of 1964), while the bottom panel shows such data over a 4-year period (1964–1968), including the information that is also shown in the top panel. Comparison of these two parts of the figure indicate that the patterns in the timing of births to teens that are salient over the 4-year time period are all but rendered invisible in the half-year time frame. The pattern shown in the bottom panel of Figure 1.2 suggests that fluctuations in the number of reported births to teens does not necessarily follow a predictable pattern over time. Over and above the dispersions over the short range, an undulating pattern can be seen over the long range of the series that is quite striking and, therefore, needs to be accounted for when we analyze these data further. Chapters 2 and 3 explain in greater detail how such an analysis might proceed.

A cautionary note is in order here. The need for information over the long range of a time series presents a major logistical challenge—a large number of successive measurements is needed to generate reliable parameter estimates. The conditions under which inferring complex processes, such as fractality, from such data is permissible has received some attention in the literature. Eke et al. (2000) tested the reliability of fractality estimates using a time series of 2^{17} ($N = 131,072$), a sample size that will not be encountered

8

Figure 1.2   Reported Number of Births to Teens in the State of Texas in Two Time Windows

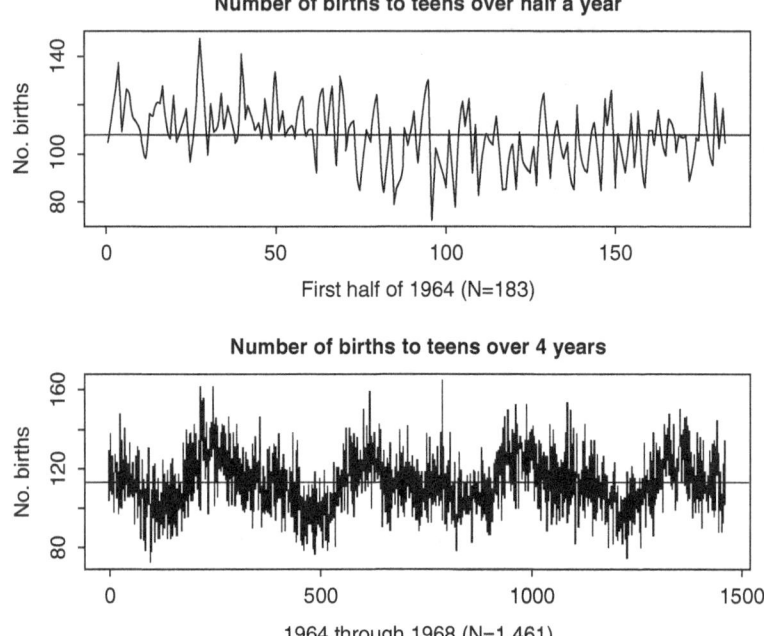

Note. From January through June 1964 (*above*); From 1964 to 1968 (*below*). The mean of the series is superimposed.

very often in the social sciences. Other researchers have considered it acceptable to conduct analyses using series of 2^9 or 2^{10} (Delignières et al., 2005). Either way, the fact remains that large numbers of sequentially ordered observations are deemed necessary to conduct the types of analyses discussed here, and many data sets where fractality is a fair question are too small to have enough resolution for its reliable detection (Keele et al., 2016). Box-Steffensmeier and Smith's (1998) data on presidential approval ratings, consumer sentiment, and macropartisanship illustrate the challenges to modeling irregularity effectively when the data sets are smaller. While the time period covered by these data is considerable (40 years), the total number of successive measurements is small ($N = 160$) relative to the power computations shown above. The question whether fluctuation patterns repeat within and across presidential terms is an intriguing one to consider, but the level of clarity and detail in the data as shown in

Figure 1.1 is insufficient to obtain any definite answers. It is therefore preferable to remain on a purely descriptive level in the analysis and interpretation of these data and draw benefit from the strength of these data, which is the fact that it provides multiple time series measuring a set of interrelated constructs, allowing for an estimation of the time dependency of their interrelationships. Meanwhile, we need further guidance from research about the extent fractal patterns can be reliably estimated in shorter time series.

Another statistical power issue of note is the inherent limitation of single-case designs and the difficulty they pose to our ability to generalize the findings to the population from which a sample was taken. Arguably, the teen births and unemployment data concern populations, but the attendance data discussed below and in subsequent chapters concern only one school, whose attendance rates are part of a huge data structure including many similar schools within the same system. Lengthy time series are often produced based on the observation of some aspect of the behavior of a single person or a unit (e.g., heartbeat). How valuable is such information in the larger scheme of things? An insightful perspective on this dilemma is offered by Molenaar (2004), who argues that we need to complement the strength of population inferences based on large cross-sectional samples with rigorously sampled data from the time spectrum through which we can describe how behavior evolves and changes, requiring the types of analyses that are the primary concern of this book.

C. The Search for Complexity

Over and above the need to model dispersion in serially ordered data as a statistical correction, the use of a long-range time perspective in such situations has also been rationalized from the vantage point of complexity theory, a set of theoretical models whose primary concern is with the adaptive behavior of systems, including human ones (Kauffman, 1995; Waldrop, 1992) and the circumstances under which transformation occurs in those systems. The applicability of such models to the social sciences has been discussed extensively in many places (e.g., Guastello, 1995; Koopmans & Stamovlasis, 2016; Sulis & Combs, 1996), attesting to the importance of time as a contributing parameter when attempts are made to understand the dynamical processes underlying the adaptation and transformation in social systems. However, such processes typically play out over the longer term and, therefore, require a long-range perspective on the data.

One behavioral manifestation of complexity that has received considerable attention in the dynamical literature is self-similarity, the replication of behavioral patterns within themselves in a fractal manner (Bak, 1996; Mandelbrot, 1997). These patterns have been said to be indicative of how systems maintain their characteristic response mode (the similarity) in the face of changing circumstances that trigger responses at variable time ranges that do not repeat in a predictable manner. Figure 1.3 illustrates this process in a large high school in New York City (here called School 1), whose attendance rates have been recorded on a daily basis (Koopmans, 2018b) for a prolonged period. The panels show those attendance rates

Figure 1.3   Evidence of Self-Similarity in Daily Attendance Rates in School 1: Patterns Repeating within Themselves across Three Different Time Scales

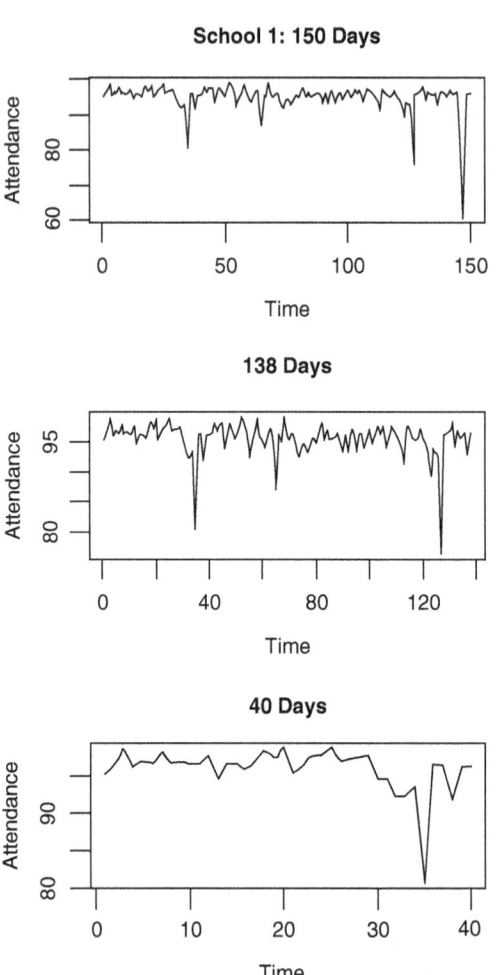

over three time chunks of varying length, with each shorter period included in the longer ones. Thus, the 150-day period shown in the top panel includes the 138 days shown in the middle panel, which in turn includes the 40 days shown at the bottom of the figure. A comparison of these three panels illustrates the influence of the size of the time window on the resulting pattern that can be seen in the figure. In Figure 1.3, these time windows have been chosen to bring out the self-similar pattern in these attendance rates, but in studies that are more confirmatory in nature, there may be a need to vary window size more systematically to determine its influence on the results.

There is considerable fluctuation in the attendance rates shown in Figure 1.3 but also a clear nonrandomness in those fluctuations. There seems to be a temporal pattern in which there is a gradual increase in variability, followed by a precipitous drop in attendance rates, and then a replication of that same process over a different time scale, suggesting a kind of tension–release pattern, in which there is a recovery of attendance after the drop and then a repetition of the cycle. The point of interest to complexity scholars in patterns such as these is that this similarity of appearance is repeated over a time scale that is not fixed. In that way, these cycles are distinct from regular cycles, such as the days of the week and months of the year, which make patterns of variability highly predictable. Arguably, the pattern shown in Figure 1.3 also has interest for educational practitioners, who may be able to informally detect cyclical (e.g., weekly) patterns in the attendance rates in their school buildings but do not have the means to evaluate irregular ones.

Another aspect of complex behavior that is of interest to many is power law distributions, distributions in which the intensity of a response is inversely related to the likelihood of its occurrence. A textbook example comes from Bak (1996), who plots the frequency with which earthquakes occur against their magnitude on the Richter scale to show the linear relationship between these two variables, such that larger quakes occur infrequently and smaller ones more often. Thus, power laws can be used to model extreme events that are not captured effectively through the use of conventional summary statistics (mean, standard deviation), which often assume a Gaussian distribution of outcomes, in which extreme observations would vanish into one of the tail ends of the curve (Adriani & McKelvey, 2007). Power law distributions represent an important alternative to the Gaussian model, and its prevalence and applicability in social science therefore needs to be better understood.

In its analysis of stability and change in systems, the dynamical literature often makes a distinction between *endogenous* and *exogenous* processes, which refer, respectively, to the prediction of a behavioral variable in terms of its own previous behavior and the prediction of a variable in terms of the behavior of other variables as they pertain to the same system. With the exception of the data of the Box-Steffensmeier and Smith (1998) study discussed above and the school size study discussed in Chapter 5, all data discussed in this book are univariate, which is to say that they are time series consisting of successive measurements on a single variable, without considering the possible influences of other variables on its behavior. In other words, the methods discussed in this book are primarily concerned with the endogenous process. The limitations of this approach are acknowledged, and much of the work presented here can be fruitfully supplemented by an analysis of the exogenous influences on the behaviors described here. Obviously, there is a trade-off between the high level of detail in a within-subject long-range time series and the analytical rigor that comes with the cross-sectional analysis of complex interrelationships between variables between subjects.

While the search for complexity often involves the detection of fractal patterns in time series, there are many other aspects to this search that are not covered in this volume. One of the most prominent alternative perspectives to complexity is the social networks perspective, which helps us better understand how different components of a system are interconnected with one another and how their connections serve to maintain a system's cohesion and ensure its effective adaptation. For the sociometric aspects of that question and its social science applications, consult Borgatti et al. (2013), Wasserman and Faust (1994), Knoke and Yang (2019), or Harris (2014). Barabási (2014) provides an excellent primer for the use of social network approaches to address complexity and dynamical processes of transformation in social and other networks. For the adaptive aspect of systems behavior, Kauffmann (1993) and Maturana and Varela (1980) are still among the best sources. Our understanding of a complex system in the social sciences ultimately requires an appreciation of the evolution of its behavior over time, its internal regulation through interactions between its constituent components, and its adaptive responses to external events and circumstances. Fractal time series, in other words, addresses but one piece of a much larger puzzle.

D. Plan of the Book

The proposed book will focus on two related and complementary approaches to the analysis of irregularity in time series. One is the *autoregressive fractionally integrated moving average* (ARFIMA) or *fractional differencing* method (Granger & Joyeux, 1980; Sowell, 1992) and the other is *power spectral density analysis* (PSDA, Delignières et al., 2005; Mandelbrot & van Ness, 1968). Fractional differencing is a more general formulation of the conventional regression-based time series model (e.g., Box & Jenkins, 1970) that has traditionally been used for forecasting—for example, in econometrics and meteorology. Fractional differencing incorporates many features of this conventional method into a larger analytical framework that is equipped to also address irregularity and long-range processes in a time series. Fractional differencing will be further discussed in Chapter 2, starting with a brief review of the basic results in time series analysis, followed by a discussion of the problem of irregularity in that context. I will then discuss how fractional differencing handles the irregular features of a time series. PSDA, discussed in Chapter 3, converts a time series into a data structure consisting of the frequencies with which cycles reoccur at given lag sizes in the series, and the contribution that these cycles make to the overall variability in the data. The PSDA approach relies on an approach called harmonic analysis (Cryer & Chan, 2008), which fits sinusoidal curves to capture cyclical trends in the series. The chapter will cover the basics of that approach and then illustrate how log transformation of these data generates the aforementioned power law distributions. It will then be discussed what these distributions tell us about irregular and fractal patterns in the data. Chapter 4 will expand on the discussion of the preceding two chapters to offer some alternate approaches to the analysis of fractal time series. Specifically, the chapter will discuss the estimation of fractal variance and the use of spectral regression to detect fractal patterns. Chapter 5 discusses some approaches to time series analysis that address related concerns such as the detection of chaos and turbulence, the prediction of time series based on other time series (i.e., multivariate time series analysis), and the analysis of long-range patterns that are regular rather than irregular, such as annual dependencies in daily recordings. Finally, Chapter 6 offers a broader perspective on the subject of the book and provides further detail about the software used to conduct the analyses.

Throughout the remainder of the book, I will rely on a set of simulations that show some of the data patterns typically encountered in these analysis and several sets of real data. The details of all these data sets can be found in the next chapter. The data sets themselves are included as text files in the appendix (available on the accompanying website for this book at **https://study.sagepub.com/researchmethods/qass/koopmans-using-time-series**), and they exemplify what time series data sets should look like. The topic of data preparation for time series analysis receives good coverage in Sayrs (1989). Most of the analyses and simulations discussed in this book are conducted using R, the open-source statistical software package. In addition, some analyses are also presented using STATA Version 14, a commercially available statistical software package. All data sets utilized in the text as well as the requisite programming statements are provided in the appendix, such that the reader can replicate each of the analyses discussed in this book. He or she also has the option of replicating the simulations or generate such simulations anew (and vary the input values if desired). Chapter 6, Section C, further comments on the use and availability of various other statistical software packages for these analyses.

2

AUTOREGRESSIVE FRACTIONALLY
INTEGRATED MOVING AVERAGE OR
FRACTIONAL DIFFERENCING

As argued in the previous chapter, the aggregation of data across the time spectrum to create snapshots of what goes on may leave important aspects of the behavior of interest unattended. This chapter discusses time series analysis as an effective way to address that problem, focusing on time series that display irregular patterns of variability across the time spectrum. The chapter starts with an overview of some of the basic insights of traditional time series analysis, and within that conceptual framework, there will be a further discussion of the use of time series analysis to estimate irregular long-range patterns in the data. An alternative approach to such data, PSDA, will be presented in Chapter 3.

A. Basic Results in Time Series Analysis

In general terms, time series analysis can serve three purposes: (1) Estimation of the impact of time, or the ordering of observations, on an outcome of interest (description); (2) enhancing the reliability of the estimation of a given outcome by modeling previous measures of this same outcome (statistical correction); and (3) projection of the behavior of a given outcome into the future based on past observations (forecasting). In each of these three cases, the main concern of the analysis is to model the statistical dependency (autocovariance, autocorrelation) in an ordered string of within-subjects observations to enhance the characterization of the distribution of time series data. Since ignoring those dependencies will yield unreliable estimates, they need to be modeled to improve our description of the data, irrespective of whether description, correction, or forecasting is done. Many introductory time series texts (e.g., Cryer & Chan, 2008; Shumway & Stoffer, 2011) discuss all three approaches. Hyndman and Athanasopoulos (2012) specifically discuss forecasting; Clarke and Lebo (2003) focus on statistical correction. The primary purpose of this book is the use of time series as a tool for statistical description.

Considering a sequence of observations $Y_1, Y_2, ..., Y_n$, one aspect of describing these dependencies is to estimate the autocorrelation function

(ACF) ρ_k for $k = 1,2,\ldots$ The sample ACF r_k over k lags can be computed as follows:

$$r_k = \frac{\sum_{t=k+1}^{n}(Y_t - \bar{Y})(Y_{t-k} - \bar{Y})}{\sum_{t=1}^{n}(Y_t - \bar{Y})^2} \quad \text{for } k = 1,2,\ldots \qquad (2.1)$$

The denominator in this formula is a sum of squared terms over the entire series, while the products in the numerator concern only $n - k$ cross-products depending on the lag(s) being estimated (Cryer & Chan, 2008).

In the estimation of the dependencies in a string of within-subject observations, time series analysis conventionally models two distinct variance components: *autoregression* and *moving average*. The *autoregression process* establishes the dependency of given observations based on the values of preceding observations. Assuming p lags, the autoregressive process, AR (p), can be defined as follows:

$$Y_t = \phi_1 Y_{t-1} + \phi_2 Y_{t-2} + \ldots + \phi_p Y_{t-p} + e_t. \qquad (2.2)$$

The number of lags included in the model can be varied to obtain the best description of the data. The *moving average process* is distinct, and it expresses the correlation between measurements in terms of their deviation from the mean of the series. Over q lags, the moving average process, MA (q) can be described as follows:

$$Y_t = \mu + e_t + \theta_1 e_{t-1} + \theta_2 e_{t-2} + \ldots + \theta_q e_{t-q}. \qquad (2.3)$$

Reduced to one lag only for each component, autoregression and moving average can be incorporated into a single model as follows:

$$Y_t = \phi_1 Y_{t-1} + e_t + \theta_1 e_{t-1}. \qquad (2.4)$$

This model is conventionally referred to as the ARMA (1, 1) model. Its more general form is ARMA (p, q), where p and q refer, respectively, to the number of AR lags and MA lags included in the estimation process.

It is good practice in time series analysis to do a preliminary exploration of the data based on a plot of the data points and an ACF plot to get a preliminary sense across the time spectrum of what goes on in the data, as well as recognizing some the patterns that are typically encountered in time series data. One important question to address is whether a time series displays random variability, or whether dependencies of the type discussed above are discernible in the series.

Figure 2.1 shows a simulation of these two scenarios, referred to here as random variability (white noise) and short-range autoregression. The panels on the left of the figure show the time series as a plot of the observed measurements Y_t against time or, to be more precise, against the order in which the measurements were taken. The panels on the right show the ACF plots of r_k versus k, which is the lag size. In these plots, the spikes represent the size of the autocorrelations at each value of k. The dotted lines in the ACF plot mark the 95% confidence interval. The two top panels in Figure 2.1 characterize a series with white noise. In such series, knowing past observations will not improve our prediction of subsequent observations as the fluctuation in the series is fully random. This simulation generated a random normal distribution with a mean of 0 and a standard deviation of 1 (see the appendix on the accompanying website for this book at **https://study.sagepub.com/researchmethods/qass/koopmans-using-time-series**). The mean in this and in the other simulations is superimposed on the time series at $Y_t = 0$. In a series in which the variability is random, one would not expect any autocorrelation between observations. Such dependencies would create an opportunity for prediction. The ACF on the top right of the figure confirms this expectation. The spikes vary randomly and generally stay within the confidence interval, and there is no real pattern to the lags at which autocorrelations are statistically significant.

Figure 2.1 A Simulation of Two Typical Data Scenarios in Temporal Data: Random Variability (White Noise) and Short-Range Autoregression

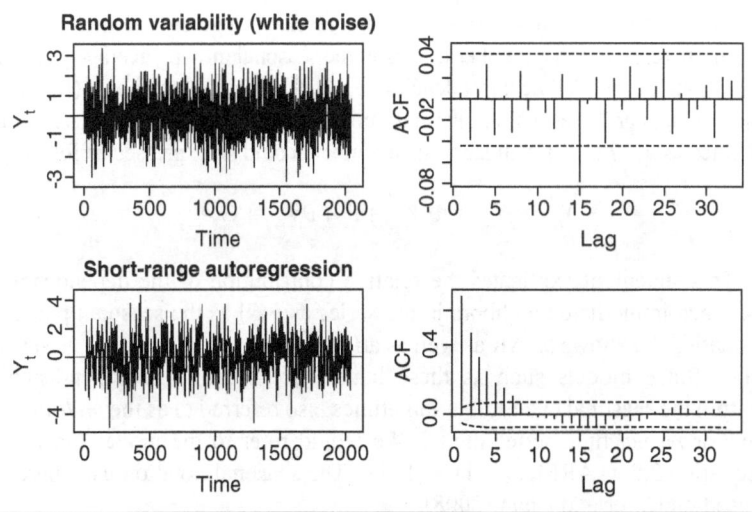

Note. Random variability (*top row*) and short range autoregression at $\varphi_1 = 0.7$ (*bottom row*). The *left panels* show time series plots; the *right panels* show the autocorrelation function (ACF) plots.

The two bottom panels in Figure 2.1 show the corresponding plots for short-range autoregression. As in the case of white noise above, a random normal distribution was specified for this simulation (see the appendix on the accompanying website for this book at **https://study.sagepub.com/research-methods/qass/koopmans-using-time-series**); the AR (1) parameter was set at $\phi = 0.7$, a fairly strong autocorrelation between immediately neighboring observations in the series. This autocorrelation can be seen in the somewhat clustery appearance of the time series plot on the left. The ACF plot on the right is also typical for a time series with short-range dependencies: While significant autocorrelations are observed at the first few lags, there is a rapid recession to statistical nonsignificance as the lag size increases. The mean of zero is superimposed on the two time series plots shown here. The appendix to this chapter (available on the website) provides further information about the input values for these simulations.

Seasonal Patterns

One important aspect in many time series analyses is the detection and modeling of seasonal regularities, such as those that might occur according to the days of the week or months of the year. Such dependencies are often substantively meaningful, and they can be readily incorporated into the short-range ARMA framework, over and above the dependencies between observations in immediate proximity. For instance, one could model dependencies according to the 7 days of the week. To estimate weekly patterns in daily measures, we can model the AR (P) and MA (Q) processes, where P (in caps) represents the number of lags used to estimate seasonal autoregression at a given periodicity, and Q similarly represents the number of lags in the seasonal moving average process. Thus, one can model the dependencies between immediate neighbors (e.g., one or two lags) and seasonal regularities simultaneously. If the seasonal period is defined as $s = 7$ (i.e., seven lags), such a model could be expressed as follows:

$$Y_t = \varphi_1 Y_{t-1} + \Phi_1 Y_{t-s} + e_t + \theta_1 e_{t-1} + \Theta_1 e_{t-s}. \qquad (2.5)$$

This statement explicates the relative contribution of the dependencies between immediate neighbors in the series as well as the seasonal pattern. Equation 2.5 shows an AR as well as an MA process at one as well as seven lags. Since models such as these handle the short-range dependencies within the seasonal cycles, it is sometimes also referred to as the *multiplicative* or *nested* time series model. We would refer to the model shown in Equation 2.5 as ARMA (1, 1) × (1, 1)$_7$. The seasonal notation used here is based on Cryer and Chan (2008).

Figure 2.2 shows a simulated example of what the time series and ACF plots would look like in a seasonal series. The figure contrasts simulations at $s = 7$ and at $s = 12$, with one seasonal parameter fixed at $\Phi = 0.4$ in both cases, but without any other short-range parameters specified in either case. Thus, the simulated models are ARMA $(0, 0) \times (1, 0)_7$ and ARMA $(0, 0) \times (1, 0)_{12}$, respectively. It can be seen in the figure that while the time series do not clearly reveal it, the seasonality in these data is clearly discernible in the spikes shown in the ACF plots.

Integration

The ARMA estimation described above assumes that the series to be analyzed has constant statistical properties, a situation also referred to as *stationarity*, or *statistical equilibrium* (Cryer & Chan, 2008). Under this assumption, the mean and variance of the series do not depend on the choice of time points within the series. One can appreciate the validity of this assumption in the plots shown in Figure 2.1, as well as to the seasonal data in Figure 2.2, in which dependencies between observations follow a predictable pattern and the mean describes the series well. Consider, instead, the time series shown

Figure 2.2 A Simulation of Seasonal Cycles at 7 and 12 Lags

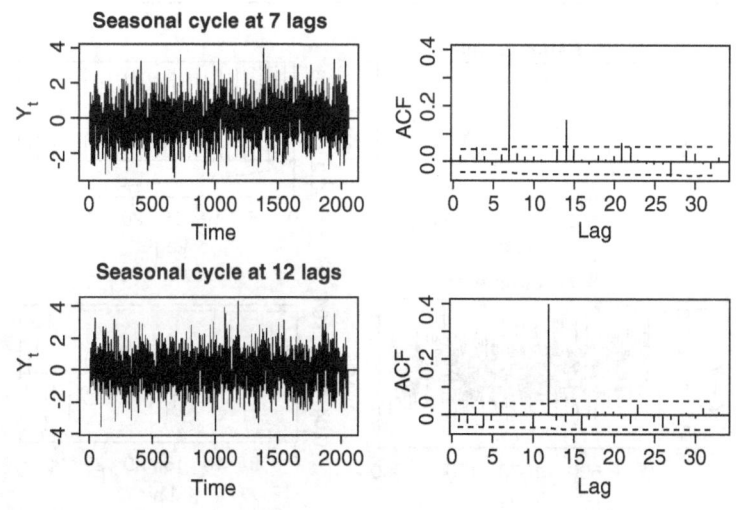

Note. Both cycles are generated at $\Phi_1 = 0.4$. The *left panels* show time series plots; the *right panels* show the autocorrelation function (ACF) plots.

in the top left panel in Figure 2.3, which shows a simulation of a pattern that is known in the time series literature as *random walk*.

Rather than having constant statistical properties, the series in Figure 2.3 can be described as volatile. It fluctuates very heavily, and the mean at $Y_t = 0$ does not provide a very good characterization of the distribution of observations. It overpredicts for long stretches of time in some places, while it underpredicts in others. Yet it can also be seen in the figure that there is a tight clustering of observations that are in close proximity. The ACF plot shown in the top right of the figure brings out this aspect more clearly. Irrespective of the lag sizes considered here, it appears that autocorrelations are highly significant and show very little recession toward nonsignificance (at least not in the time window considered here).

A random walk presumes observations that are discrete along the time scale. Of theoretical interest in such cases is often an underlying continuous pattern of randomness called *Brownian motion*, so named after the British botanist Robert Brown who, in the early 19th century, detected and described random fluctuation in the behavior of pollen seen under a microscope (Feder, 1988). However, in our modeling efforts, we often interpret data showing random walk in terms of such an underlying pattern even though the approximation of it is achieved by a set of discrete

Figure 2.3   A Simulated Random Walk and its First Difference

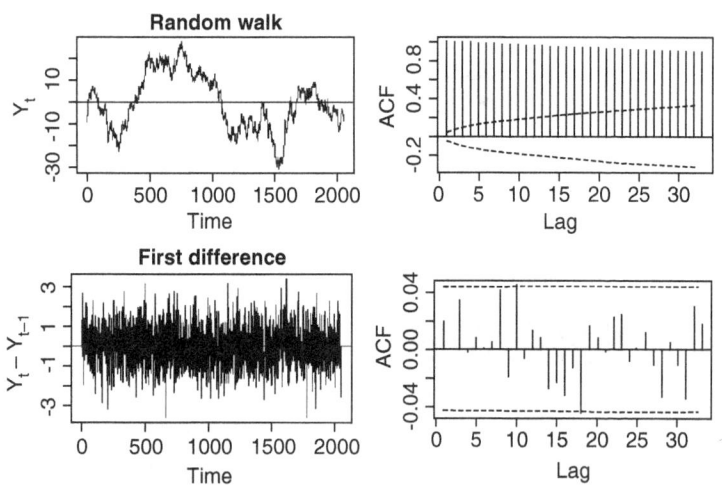

Note. The random walk (*top row*) is the cumulative sum of white noise. The first difference (*bottom row*) shows its stationary increments. *Left panels*: Time series; *Right panels*: Autocorrelation Function (ACF) plots.

measurements (i.e., the "steps" in a random walk). It is important to keep in mind that Brownian motion and random walk are distinct statistical constructs, although both would manifest the type of nonstationarity shown in Figure 2.3.

The conventional response to nonstationarity in a time series analysis is to attain stationarity through differencing, typically (but not necessarily) the first difference $\nabla Y_t = Y_t - Y_{t-1}$, which is intended to produce a series consisting of *stationary increments*. Cryer and Chan (2008) provide a more general formulation of the integration process. A time series Y_t follows an autoregressive integrated moving average (ARIMA) pattern if the d-th difference $W_t = \nabla^d Y_t$ is a stationary ARMA process, in which case, Y_t is said to be an ARIMA (p, d, q) process, with no differencing undertaken if $d = 0$, the first difference being $d = 1$, and so on. Hence, an integrated model in which autoregression and the moving average processes are estimated at one lag only, would be called an ARIMA (1, 1, 1) process, whereas if no integration is undertaken but the same autoregressive parameters are included in the model, we call it ARIMA (1, 0, 1).

The two bottom panels of Figure 2.3 show the effect of taking the first difference $d = 1$ on a simulated random walk. The time series plot on the bottom right shows how the first difference (left column) removes the volatility and produces stationary random variability instead. The corresponding ACF plot on the right confirms this impression by showing a random pattern instead of the heavily autocorrelated pattern shown in the top right panel.

Testing for Stationarity

To confirm our impressions about the (non)stationarity of time series data, many statistical tests are available. Among the most well-known are the Augmented Dickey–Fuller (ADF) unit root test (Said & Dickey, 1984) and the Kwiatkowski–Phillips–Schmidt–Shin (KPSS) test (Kwiatkowski et al., 1992). The basic idea behind the ADF test is that an actual series is regressed on itself at a given number of previous lags, as well as on a stationary series, such that it can be estimated whether model improvement is attained by adding the stationary process. The null hypothesis in the ADF test is *nonstationarity*.

Based on the argument that rejecting the alternative hypothesis in an ADF test does not necessarily imply the types of nonstationarity assumed under a random walk or Brownian motion, the KPSS test is concerned instead with the status of the null hypothesis of *stationarity*, specifically tested against a deterministic trend with a nonzero slope and against a random walk. The KPSS test makes an explicit distinction, then, between the test for a linear trend—that is, if the series moves upward or downward as time progresses—and nonstationary error, which corresponds to random walk. The distinction between these two types of nonstationarity can also be made when performing an ADF test, but less easily so (see Cryer & Chan, 2008). Given the difference between these two approaches (testing for nonstationarity vs. testing for stationarity), Kwiatkowski et al. (1992) suggest that the two tests be used in a complementary fashion, a recommendation that is followed in this volume. Stationarity testing constitutes a first step in a longer process to remove nonrandomness from the data through a systematic modeling process. The following section describes this process in greater detail.

B. Long-Range Dependencies

One might imagine time series that are not quite as volatile as those shown in Figure 2.3, while also not displaying the degree of stability shown in Figures 2.1 and 2.2. Fractional differencing has been developed to specifically address this intermediate situation, which is sometimes referred to as *metastability* in recognition of the fact that the series could be not quite stable and not yet volatile either, or *pink noise*, not quite white noise where all frequencies have an equal probability, but varying probabilities across the series. In the dynamical literature, this pattern is also sometimes referred to as $1/f$ noise (see Chapter 3, Equation 3.9, for further context). Models are needed that enable us to characterize this intermediate state in a time series. Fractional differencing proposes a solution that uses neither $d = 0$ for stationary series nor $d = 1, 2, \ldots$ for nonstationary ones but, instead, estimates d as a fraction that is assumed to be within the range of $-0.5 < d < 0.5$ to indicate the extent to which metastability is found in the series. This estimation involves a statistical test of whether d is different from zero. A differencing parameter greater than zero indicates the presence of positive autocorrelations over the long range of the series, also referred to as *persistence*. If $d < 0$, there are negative autocorrelations over the long range, a process also called *antipersistence*.

To explicate the fractional differencing model, it is helpful to use a statistical convention called the *characteristic equation*, which in turn relies on a mathematical function object called the *backshift operator* or the *lag operator*. Given a time series $Y_1, Y_2, \ldots, Y_n$, the backshift operator is defined as

$$BY_t = Y_{t-1}. \tag{2.6}$$

And likewise,

$$BBY_t = B_t^2 Y_t = Y_{t-2}. \qquad (2.7)$$

The higher order terms in this formulation, then, represent the diminishing influence of past observations on the series as lag size increases. The general model incorporating both differenced and fractionally differenced solutions is sometimes called ARFIMA (p, d, q) or the autoregressive fractionally integrated moving average model, and it can be written as follows:

$$\left(1 + \varphi_p B^p\right)(1 - B)^d Y_t = \left(1 + \theta_q B^q\right) e_t. \qquad (2.8)$$

In this equation, the leftmost term represents the autoregressive component to the variability in the data and the second term on the left represents the differencing component. It can be readily appreciated that in the nondifferenced case ($d = 0$), the term in the middle disappears, while if there is differencing (e.g., $d = 1$), it shows the first, second, or n th difference of the series. The estimation of d is used to establish a pink noise pattern in the data. The right-hand side of the equation shows the moving average component. As is common in probability statistics, the modeling process assumes an independent identically and normally distributed error term; that is,

$$e_t\left(t = 1, 2, \ldots\right) \sim N\left(0, \sigma^2\right) \text{ IID}, \qquad (2.9)$$

where IID = independent and identically distributed.

Figure 2.4 shows the time series and ACF plots for a simulation of pink noise at $d = 0.35$, a series in other words with quite a bit of persistence in it. As is the case in short-range autoregression, there tends to be a clustering of observations, but even more so in this case and over the longer range of the time scale, resulting in a kind of undulating pattern, which is in fact quite characteristic of pink noise. Likewise, the ACF plot for pink noise shows a slow recession toward nonsignificance of the autocorrelation as the lag size gets higher, and while this recession is slower than in the case of short-range autoregression, it is more decisive than in the case of the random walk shown in Figure 2.3.

The time series literature provides a helpful metaphor for the characterization of the four data scenarios discussed in these two sections, which is in terms of the *memory* in the series, respectively, as follows: no memory (white noise), short-range memory (short-range autoregression), long-range memory (pink noise), and infinite memory (random walk, or Brownian motion, see, e.g., Stadnitski, 2012a). Another comparative look

24

Figure 2.4   A Simulation of Metastability (Pink Noise)

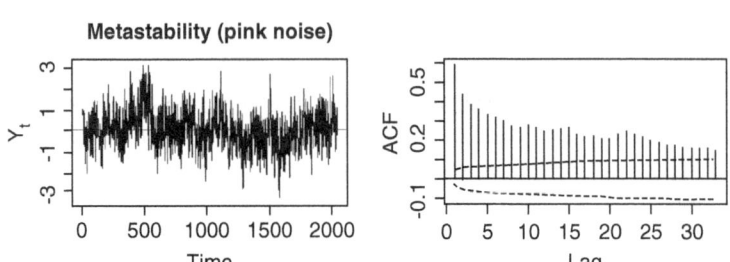

Note. Time series (*left panel*); the autocorrelation function (ACF) plot (*right panel*). This pattern is generated at $d = 0.35$.

at the ACF plots for these four scenarios should help appreciate the idea of memory in the time series context.

C. Application of the Models to Real Data

In this section, I will demonstrate how the trends discussed above show up in real time series data and how time series analysis and fractional differencing can be used to seek confirmation for them. In this section, and subsequent chapters of the book, I will rely on the following four data sets:

1. *Monthly unemployment figures in the United States* from 1948 through 2017 ($N = 837$). These statistics are available from the U.S. Department of Labor (2017). The data are based on the Current Population Survey, a monthly survey conducted by the Census Bureau for the U.S. Department of Labor.

2. *Daily attendance rates in one urban high school from 2010 through 2014* ($N = 735$), vacations excluded (Koopmans, 2015). These data are based on daily school attendance records accumulated by the New York City Department of Education. Extreme observations were replaced in this data set by a linear combination of neighboring observations, creating an "uncontaminated series."

3. *The daily number of births to teens in the state of Texas from 1964 through 1990* (Hamilton et al., 1997). The data were obtained from the Texas Department of State Health Services, and they were derived from birth certificates. The full data set is quite sizable ($N = 13,149$), which makes it particularly suitable for the analysis of long-range processes. A smaller fragment is used in the examples in this chapter, namely, births from January 1, 1964, through March 10, 1966 ($N = 800$).

4. *Left–right political orientation in the Netherlands* as measured by an opinion survey administered weekly from January 1978 through December 1996 ($N = 988$) to a random sample of respondents, with findings disaggregated by declared party affiliation (Eisinga et al., 1999).

These four data sets were chosen because of their suitability to illustrate different aspects of the dependency patterns described above and due to their relevance to social science research.

Figure 2.5 shows the time series and ACF plots for three of the four data sets introduced above. The fourth one, left–right political orientation in the Netherlands, is discussed in detail later in the chapter. Understandably, as the sample size of the time series data sets get larger, the range of the confidence intervals in the ACF plots gets smaller. Therefore, only a fragment of the attendance and births to teens data ($N = 800$) is shown in these figures, so that the sample size in the three data sets is approximately the same and the dependency patterns shown in the ACF plots can be evaluated against confidence intervals of comparable size.

A few patterns are noteworthy in these three data sets. The daily attendance rates show some tight clustering as well as quite a few extreme values pulling the daily attendance downward. These outlying values have been found to be associated with unusual circumstances such as inclement weather and upcoming holidays (Koopmans, 2015). The ACF plots for daily attendance show a recession toward statistical nonsignificance but also peaks at the fifth lag and multiples thereof, suggesting a seasonal pattern corresponding to the days of the school week. The unemployment numbers in the second row of the figure show a typical example of a nonstationary series, with the tight clustering of measurements that are close together, a high degree of volatility over the long range of the series, and an infinitely slow recession of the autocorrelations toward nonsignificance. The effect of taking the first difference on the unemployment data is shown in the third row of the figure. While doing so clearly addresses the volatility in the data, the clustery pattern between observations in close proximity that characterized the original series can be seen here as well, both in the time series and ACF plots, in which the spikes representing autocorrelations between observations in close proximity are significantly different from zero, and a spike at the 24th lag suggests a seasonal pattern. This result indicates that nonrandomness remains in the data that will need to be addressed in subsequent analysis.

In the last row of Figure 2.5, births to teens in Texas are shown for a period of 2 years and somewhat over 2 months. The time series shows the undulating patterns suggestive of pink noise. The slow regression toward nonsignificance is consistent with this impression. However, as in the school attendance data, the ACF plot also shows a pronounced seasonal pattern according to the days of the week (which in this case includes the

26

Figure 2.5 Daily School Attendance, U. S. Unemployment Numbers and Births to Teens in Texas: Time Series and Autocorrelation Function (ACF) Plots

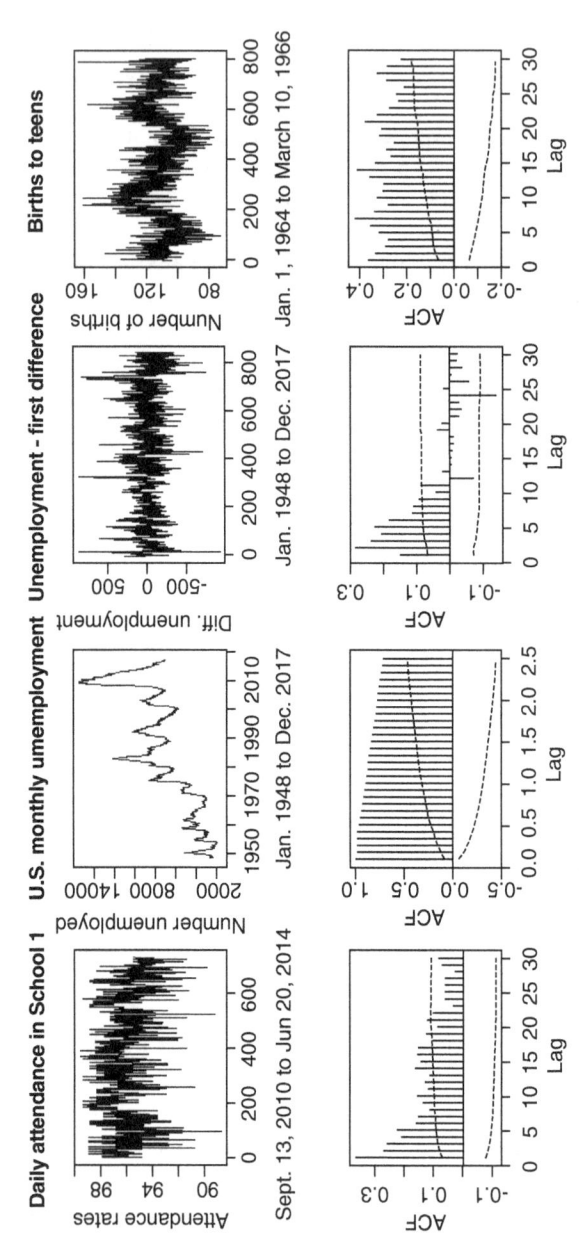

Note. Time series plots (*top row*) and ACF plots (*bottom row*), showing daily attendance rates in School 1 from September 13, 2010 through June 20, 2014 (*first column*), U.S. monthly unemployment figures from January 1948 through December 2017 (*second column*), the first difference of the U.S. unemployment figures (*third column*) and daily number of births to teens in the state of Texas from January 1, 1964 through March 10, 1966 (*last column*). Outliers were removed from the attendance series.

weekends). To estimate the relative significance of the metastable and seasonal patterns in these data, it will be necessary to analyze these data statistically, as will be done in the section that follows. The plots in Figure 2.5 seem to suggest a metastable pattern in the Texas births to teens data and possibly also in the daily attendance rates in School 1. The time series and ACF plots are not conclusive by themselves, and hence, additional statistical modeling is required to corroborate these impressions.

As a first step in this process, stationarity tests need to be conducted to determine what if any differencing strategy needs to be used to address instability in these data. Table 2.1 shows the results of the ADF and KPSS tests for the daily attendance rates in School 1, the U.S. unemployment data, the births to teens

Table 2.1 Results for the Augmented Dickey–Fuller (ADF) Unit Root Test for Nonstationarity and the Kwiatkowski–Phillips–Schmidt–Shin (KPSS) Test for Stationarity by Data Type and Data Source: Simulated and Original Data

Data Type and Source	ADF	KPSS: Level	KPSS: Trend
Simulated data			
White noise	−12.78*	0.17	0.02
Short-range autoregression	−11.64*	0.07	0.07
Pink noise	−7.20*	2.06*	0.50*
Random walk	−2.07	3.34*	1.66*
Seasonal at 7 lags	−9.36*	1.96	0.17
Seasonal at 12 lags	−8.42*	0.12	0.05
Original data			
Daily attendance in School 1	−7.11*	1.22*	0.31*
U.S. unemployment data	−3.61	7.85*	0.34*
Births to teens data	−8.25*	24.23*	1.06*
Left–right orientation (CDA)	−5.59*	11.52*	0.44*
Left–right orientation (PvdA)	−4.91*	8.93*	0.28*

Note. Lag order for ADF tests: $(n - 1)^{1/3}$; Lag order for KPSS test: $3 \times \sqrt{n} / 13$. CDA = Christen-Democratisch Appèl (Christian Democratic Appeal); PvdA = Partij van de Arbeid (Labor Party).

*$p < .01$.

28

Figure 2.6 Left-right Political Orientation in the Netherlands from 1978 through 1996 by Party Affiliation: Original and Detrended Series

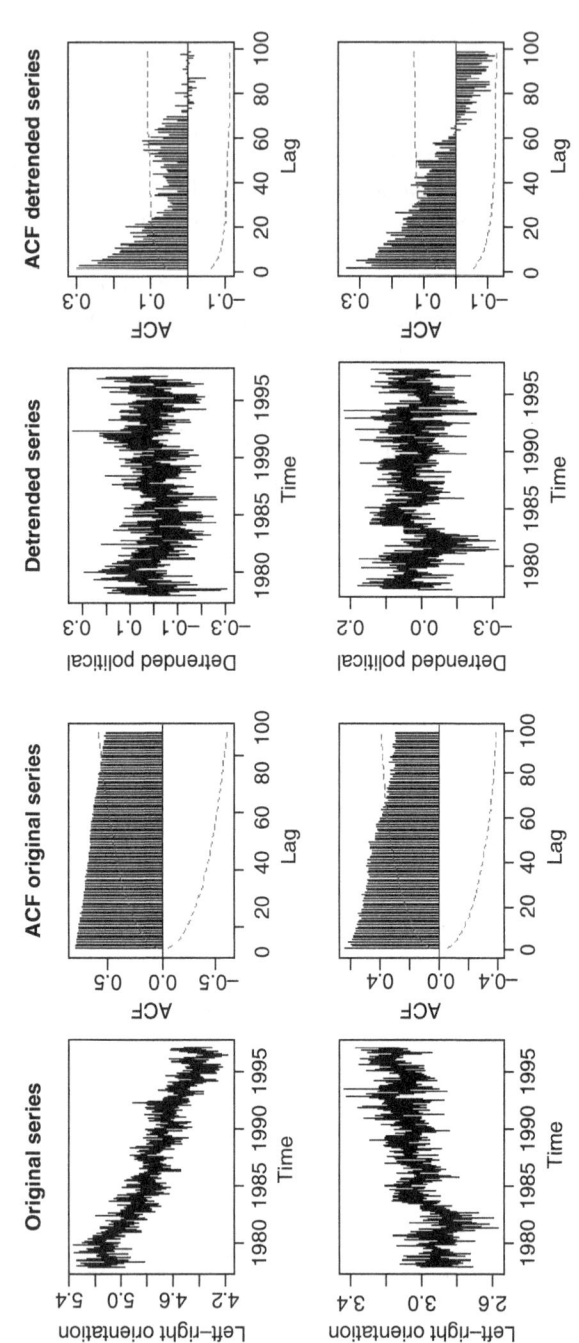

Source. Adapted from Eisinga et al. (1999).

Note. Respondents claiming affiliation with the CDA (*top row*) and with the PvdA (*bottom row*). CDA = Christen-Democatisch Appèl (Christian Democratic Appeal); PvdA = Partij van de Arbeid (Labor Party).

data, and the left–right orientation by political party affiliation in the Nether-lands. Acronyms for the parties are explained in the footnote to Table 2.1 and in the description of Figure 2.6. For comparative purposes, the stationarity test results for the simulated data discussed in the previous section are provided as well. Starting with those, the ADF test results would lead one to reject the nonstationarity hypothesis in the cases of white noise, short-range autoregres-sion, and pink noise, while in case of the random walk, there is not enough evidence to reject this hypothesis. This result is consistent with expectations. The KPSS test shows that there is not enough evidence to reject the hypothesis of stationarity in the cases of white noise and short-range autoregression. It can also be seen in the table that the two simulations of seasonal cycles behave as stationary series on these tests. The stationarity hypothesis is not confirmed, however, in the pink noise and random walk cases. The equivocal results when comparing the results of the ADF and KPSS tests in the case of pink noise points to the possibility of a metastable pattern. With respect to the "real" data discussed here, it can be seen that the ADF test rejects nonstationarity for the attendance, births to teens, and left–right orientation data, but not for the unem-ployment data. The KPSS test rejects the stationarity hypothesis in all of these cases. Taken together, the findings for the ADF and KPSS tests on the real data indicate that further testing for metastability is of interest in all cases but the unemployment data, where a series of the first- or perhaps second-order differ-ence should be the basis of any subsequent modeling efforts.

Establishing whether or not a time series is stationary is a critical step in any time series analysis because the stationarity assumption underlies the use of ARIMA models. Based on test results reported here, as well as the visual inspection of the plots, the possibility of metastability or pink noise needs to be further explored in the daily attendance, births to teens data, and left–right political orientation for both of the political parties shown in the table. We will do so in the next section, where we will also show how to proceed with the analysis of nonstationary patterns through differencing, using the unemployment data as an example.

Table 2.2 Fractional Differencing Estimates for the Births to Teens and School Attendance Data: d Coefficient and Error Variance

Data Source	d	σ^2
Births to teens data ($N = 800$)	0.26*	160.78
Daily attendance rates in School 1 ($N = 735$)	0.27*	2.81

*$p < .05$.

Table 2.2 shows the output for the fractional differencing analysis for the births to teens data (N = 800) and for daily attendance rates in School 1 (N = 735). For the births to teens, d = 0.26, which is significantly different from zero. For the daily attendance rates in School 1, d = 0.27, also different from zero. In the models discussed here, only the differencing parameter is estimated. Hence, we would refer to these models as fractionally differenced ARIMA (0, d, 0). The results shown here indicate persistence in both of these time series, making it worth our while to look for signs of self-similarity, as was done for School 1 in Figure 1.3 of the previous chapter.

Competitive Modeling Strategies

It is well-known that the estimation of short-range, long-range, and seasonal dependencies tends to yield correlated predictors. It is therefore not sufficient to simply establish that a differencing or seasonal parameter is different from zero without looking at whether the parameter would remain significant if other predictors are added to the model. Hence, it is essential to find out how much reduction there is in the variability with each term added to the model (Wagenmakers et al., 2004). This strategy is akin to stepwise regression, and it typically involves a determination of the influence of short-range autocorrelations first (e.g., at one or two lags), and the addition of either seasonal or differencing parameters, or both, and comparing the goodness of fit of the models with and without those additions. The approach should be guided by the visual impressions from the plots, and it allows one to determine the effect of differencing over and above that of seasonal and short-range estimates, as will be illustrated below.

When comparing models, there are two aspects to the determination of goodness of fit in a time series analysis, which are related but distinct. The first aspect is *maximizing the reduction of variance*. The second one is *effective removal of autocorrelation*. The two most commonly used goodness-of-fit measures for variance reduction in ARIMA-based time series analysis are Akaike's information criterion (AIC) and the Bayesian information criterion (BIC). Both are based on the log likelihood (LL) function (Cryer & Chan, 2008; Eliason, 1993) and can be computed as follows: $-2LL + 2k$ (AIC) and $-2LL + k \log(n)$ (BIC). In these two equations, k is the number of parameters estimated by the model (i.e., $p + d + q + 1$, if the intercept is included in the estimation and $p + d + q$ otherwise) and n is the number of observations in the time series. While both measures penalize for the addition of parameter estimates to the model, BIC is the more conservative of the two and shows less sensitivity to increases in sample size. Based on preliminary data, Wang and Liu (2006) suggest that BIC may be the better goodness-of-fit measure in case of nested seasonal models. Thus, since many of the analyses presented here compare nested and nonnested models, BIC will be used throughout.

In terms of the removal of autocorrelation in the data, one of the most well-known approaches is the Ljung–Box Portmanteau (LBQ) test, which evaluates the joint magnitude of the residual covariances r^2 over k lags for a given model using the following test statistic:

$$Q = n(n+2)\left(\frac{\widehat{r_1^2}}{n-1} + \frac{\widehat{r_2^2}}{n-2} + \ldots + \frac{\widehat{r_k^2}}{n-k}\right). \tag{2.10}$$

Q is tested under a χ^2 distribution with $k-1$ degrees of freedom (Cryer & Chan, 2008). If Q is significantly different from zero, we conclude that autocorrelation remains and further modeling is required. An important difference between AIC and BIC on the one hand and LBQ on the other is that the former two measures are based on a ratio of the variability accounted for by the model relative to the variability due to error, while LBQ is specifically concerned with autocorrelation not accounted for by the model. In practice, the two types of information can be used as complementary indices.

Recall the pattern shown in Figure 2.5 for the unemployment data whose first difference is taken. Inspection of the time series plot suggests a clustery pattern. The ACF plot at the bottom right of the figure confirms this impression by showing statically significant autocorrelations between neighboring observations. The plot also shows that at 24 lags, a significant autocorrelation is observed as well. Given these plots and the results of the stationarity tests reported in Table 2.1, the analysis will proceed based on a differenced series $d = 1$ and estimate patterns of dependency in the stationary increments shown in the figure. Thus, since the differencing parameter is fixed, the remaining concern is with the relative importance of the short-range and seasonal dependencies, which can be decided using the comparative modeling approach outlined above. For the modeling of the seasonal components, the ACF plot suggests that there might be an annual dependency in the data, which will be modeled here at the 12 lags that constitute the months in a year (rather than the 24 lags in the ACF plot). It will be easier to interpret an annual cycle in these data, and it is possible that modeling the annual cycle may leave us without any remaining dependencies in the data.

Table 2.3 shows the output of this analysis. The second column of the table shows the conventional ARIMA specification for the three models that are compared. Model I is an unconditional model, which quantifies the variability around a mean of zero without further specification. Model II includes one autoregressive as well as one moving average parameter estimate, ϕ_1 and θ_1, respectively. It can be seen that these parameters are both significantly different from zero, under a normal distribution in a two-sided test. Both estimates in Model II estimate the dependency at one lag, which is to say that the dependency is modeled between a given

Table 2.3 Statistical Models of Differenced Unemployment Numbers With and Without an Annual Seasonal Component (12 lags): Comparative Model Selection, Parameter Estimates, and Goodness-of-Fit Statistics

Model	ARIMA (p, d, q) Specification	Parameter Estimates					Goodness of Fit			
		φ_1	θ_1	θ_2	Φ_1	Θ_1	σ^2	BIC	2δBIC[a]	LBQ
I	$(0, 1, 0)$	—	—		—	—	42,480	11,288.27	—	283.37*
II	$(1, 1, 1)$	0.91*	−0.75*		—	—	37,526	11,198.33	89.94*	49.73*
III	$(1, 1, 2) \times (1, 0, 1)_{12}$	0.91*	−0.90*	0.18*	0.53*	−0.80*	33.175	11,095.78	102.55*	9.86

Note: ARIMA = autoregressive integrated moving average; BIC = Bayesian information criterion; LBQ = Ljung–Box Portmanteau.

[a]2δBIC is tested under a χ^2 distribution. Degrees of freedom (*df*) are decided by the number of parameters added or removed in the model comparison. Thus, *df* = 2 in the comparison between Model I and Model II, and *df* = 3 in the comparison between Models II and III. The associated critical values are $\chi^2 = 5.99$ and $\chi^2 = 7.82$, respectively.

*p < .05.

observation and its immediately preceding neighbor (autoregression) and its error term (moving average). Model III adds a seasonal component at 12 lags, the number of months in a year (and a two-lag moving average parameter). Both a seasonal autoregressive and a moving average term are included in the model, typically depicted in capital Greek letters as Φ_1 and Θ_1, respectively, at 12 lags. Table 2.3 shows that both seasonal terms are statistically significant, over and above the statistical significance of ϕ_1, θ_1, and θ_2.

As discussed above, since the comparison involves a nested seasonal model (Model III), the BIC is used as a goodness-of-fit measure. The two additional indicators of the effectiveness of the models describing the data included in the table are the residual variance σ^2 and the LBQ test. Comparing these measures indicates that both the second and the third models, but especially the third one, constitute a significant improvement. The residual variance as well as the BIC get reduced substantially in each of the two modeling steps. The LBQ test shows that it requires the inclusion of the seasonal component in Model III to reduce the value of the Q statistic to nonsignificance, indicating the successful removal of autocorrelation in the series. Based on this information, we conclude that Model III provides the best description of the unemployment data and that from 1948 through 2014, there is a monthly seasonal cycle to unemployment in the United States over and above the volatile pattern that is on display in the original series and the dependency between immediately neighboring observations.

Due to their inclination to be correlated, short-range, long-range, and seasonal estimates often mimic one another when their relative contribution is evaluated to the reduction of variance due to modeling. The significance of fractional differencing needs to be estimated very carefully to ensure that variability attributed to long-range dependencies cannot also be accounted for by the short-range estimates. Wagenmakers et al. (2004) suggest using a statistical test of the difference between goodness-of-fit indicators across models to estimate whether model improvements due to inclusion of a differencing parameter are significantly different from zero, indicating that this parameter uniquely contributes to a reduction of variance in the data. The test statistic is twice the difference between the goodness-of-fit measures, tested under a χ^2 distribution, with the difference between the number of parameter estimates in the two models as degrees of freedom. Wagenmakers et al. discuss the use of this approach comparing AIC across models; Wang and Liu (2006) use it to compare log likelihood ratios. The approach is extended here to the comparison of BIC indicators. The statistical significance of the values of twice the difference in BIC indicators indicate that Model II performs better than Model I and Model III performs

Table 2.4 Statistical Models of Daily Attendance Rates in School 1: Comparative Model Selection, Parameter Estimates, and Goodness-of-Fit Statistics

Model	ARIMA (p, d, q) Specification	Parameter Estimates						Goodness of Fit			
		Inter-cept	φ_1	θ_1	Φ_1	Θ_1	d	σ^2	BIC	2δBIC with Model II[a]	LBQ
I	$(0, 0, 0)$	96.09*	—	—	—	—	—	3.45	2,998.87	294.28*	546.96*
II	$(0, d, 0)$	96.02*	—	—	—	—	0.27*	2.81	2,851.73	—	6.65
III	$(1, 0, 1)$	96.08*	0.87*	−0.65*	—	—	—	2.83	2,858.60	13.74*	7.41
IV	$(1, d, 1)$	96.00*	0.26*	−0.32*	—	—	0.31*	2.81	2,859.47	15.48*	6.64
V	$(1, 0, 1) \times (1, 0, 1)_5$	90.04*	0.84*	−0.61*	0.97*	−0.94*	—	2.81	2,860.33	34.40*	3.95

Note. ARIMA = autoregressive integrated moving average; BIC = Bayesian information criterion; LBQ = Ljung–Box Portmanteau.

[a]2δBIC is tested under a χ^2 distribution. Degrees of freedom (df) are decided by the number of parameters added or removed in the model comparison. Thus, $df = 1$ in the comparisons between Models I and II and between Models III and IV; $df = 2$ in the comparison between Models IV and II; $df = 3$ in the comparison between Models II and III; and $df = 5$ in the comparison between Models II and V. The critical values, respectively, are $\chi^2 = 3.84$ $(df = 1)$, $\chi^2 = 5.99$ $(df = 2)$, $\chi^2 = 7.82$ $(df = 3)$, and $\chi^2 = 11.07$ $(df = 5)$.

*$p < .05$.

better than Model II. Consistent with the findings of the LBQ test, then, BIC values favor Model III, indicating that it is most effective in reducing the overall variance in the data.

Table 2.4 compares five ARIMA models to describe the daily attendance rates in School 1. You may recall that the ADF test shown in Table 2.1 rejects stationarity in these data, while the results of the KPSS test reject nonstationarity. The estimation of the independent contribution of the differencing parameter to the variability in daily attendance rates is therefore a pertinent issue. The ACF plots of the daily attendance trajectory in Figure 2.1 point to short-range autocorrelation (i.e., rapidly declining value of the autocorrelations at the first few lags. Moreover, the spikes at the 5th and 15th lag suggest that there may be a seasonal factor at work (i.e., 5 days in the school week). In addition, the slow recession of the autocorrelations to nonsignificance overall is indicative of pink noise. Given that there are pointers to all three of these processes in the plots, the comparative modeling approach enables the investigator to sort out the relative contribution of each to the overall variability in the data to then decide on the best model.

Table 2.4 summarizes the results of this analysis for the attendance data. Model I estimates the intercept only; Model II adds the differencing parameter. Comparison of Models I and II estimates the contribution of the differencing parameter with no other parameters in the model. Models III and IV perform the same assessment, but with one autoregressive and one moving average parameter added to both models. Last, Model V adds to these latter two parameters a seasonal component at five lags, the days of the school week.

The models shown in Table 2.4 list the parameter estimates for each model, as well as the goodness-of-fit statistics. Working our way from the right to the left, the results of the LBQ test indicate that Models II, III, IV, and V are all acceptable models, because none of the Q statistics are significantly different from zero. These models, then, leave no significant remaining autocorrelation in the data. The residual variance estimates (σ^2) for these four models are consistent with this finding as the differences between Models II, III, IV, and V are virtually indistinguishable. The BIC values for the five models indicate that Model II, which includes the differencing parameter but no other estimates, has the lowest BIC value at 2,851.73. The BIC values for the other models are higher. The penultimate column shows the results of the statistical test of the difference of the BIC values across models, comparing Models I, III, IV, and V with Model II. Since the difference between BIC values are statistically significant in all four cases, we conclude Model II is superior to the other three models.

It is instructive to review the parameter estimates of these models as well. Model II shows that there is considerable persistence in these data at $d = 0.27$, yet the seasonal estimates generated by Model V are highly

Table 2.5 Comparative Modeling of a Sample of the Births to Teens Data (N = 800): Fixed and Estimated Differencing Parameters and Goodness-of-Fit Statistics

Model	Specification	d	σ^2	BIC	LBQ
I	(0, d, 0)	0.26*	160.79	6,336.90	47.26*
II	(1, d, 1)	0.49*	154.60	6,314.71	31.35*
III	(1, 1, 1)	1.00	152.50	6,291.89	34.48*
IV	(1, 1, 1) × (0, 0, 1)$_7$	1.00	150.00	6,281.67	13.72
V	(1, 1, 1) × (1, 0, 1)$_7$	1.00	136.20	6,214.47	11.38

Note. BIC = Bayesian information criterion; LBQ = Ljung–Box Portmanteau.
*p < .05.

significant as well, as are the short-range parameters estimated in Models III through V. These findings are all consistent with the visual impressions shown in Figure 2.1, which shows all three patterns. BIC includes a penalty for adding terms to the model, and with respect to these findings, it is therefore right to say that Model II owes its effectiveness to its parsimony: The long-range estimation only suffices to account for the variability in these data, of which pink noise is the best characterization since it absorbs these other patterns as well.

Table 2.5 shows five possible models to describe the variability in births to teen counts in Texas from January 1, 1964, through March 10, 1966. Models I and II estimate the differencing parameter, with and without addition of short-range estimates at the first lag. It can be seen that both models indicate metastability; in Model II, the 0.49 value suggests a pattern close to volatility. However, consideration of Q statistics for the five models indicate that it takes a seasonal moving average estimate (7 days) at one lag to characterize the data effectively. Both Models IV and V are seasonal, and they yield favorable goodness-of-fit results. The overall variance is reduced most effectively by Model V.

There is a trial-and-error aspect to model fitting of this kind. For instance, the investigator might consider adding terms, such as perhaps ϕ_2 and/or θ_2, or additional seasonal estimates to the models shown in Table 2.5 to determine whether additional model improvement can be obtained. (I leave it up to the reader to evaluate this possibility.) It is therefore instructive to examine the residuals of the best fitting models to confirm that our modeling efforts have been successful, as will be done further below.

Differencing Detrended Data

One important aspect of nonstationarity that deserves separate attention is the possibility of an upward or downward trend in the data that is not necessarily accompanied by the random walk characteristic of Brownian motion. Figure 2.6 shows an example of such data, which are based on weekly surveys that have been conducted in the Netherlands from 1978 through 1996 ($N = 988$). The purpose of the surveys was to assess the political left to right orientation over time among those attesting their affiliation to particular political parties (Eisinga et al., 1999). The study notes a convergence toward the center during this period, a finding that begs the question what processes generate this trend. To address one aspect of this question, the study examines the persistence in the data trajectories for the six major political parties in the Netherlands. Figure 2.6 shows the results for two of these, the Christen-Democratisch Appèl (CDA, or Christian Democratic Appeal), a party toward the conservative end of the political spectrum, and the Partij van de Arbeid (PvdA, or Labor Party), which leans toward the left.

Comparison of the original series of these two parties shows a depolarization in left–right orientation of survey respondents over time in both groups, with a pronounced shift toward the left in the declared CDA affiliates and toward the right in the PvdA affiliates. The ACF plots of the original two series, shown here at 100 lags, show the "infinitely" slow regression of the autocorrelations toward nonsignificance, attesting to the nonstationarity of the series. Yet these ACF plots do not reveal that the nonstationarity is of a very different character here from the volatility in the employment data discussed above, and the results of the ADF and KPSS stationarity tests shown for these data in Table 2.1 are more characteristic of a pink noise process than of Brownian motion. Detrending the data enables the investigator to assess whether there is metastability or persistence in these data, a concern that is also addressed in the original study.

The model being fitted to these data is a variation to the characteristic equation (Equation 2.8) and can be formulated as follows:

$$\left(1 + \varphi_p B^p\right)(1 - B)^d \left(Y_t - \beta_0 - \beta_1 t\right) = \left(1 + \theta_q B^q\right) e_t. \tag{2.11}$$

The linear regression model at the center of this equation performs the correction needed to remove the linear trend from these data (Eisinga et al., 1999), resulting in the detrended series and its corresponding ACF plots shown in the two rightmost columns in Figure 2.6. The patterns shown in the ACF plots point to the possibility that the left–right orientation may be metastable in both affiliate groups, as the recession to statistical

Table 2.6 Best Fitting Models for Detrended Left to Right Orientation by Party Affiliation (PvdA vs. CDA): Parameter Estimates and Goodness-of-Fit Statistics

Party Affiliation	ARIMA (p, d, q) Specification	Parameter Estimates			Goodness of Fit		
		φ_1	θ_1	d	σ^2	BIC	LBQ
PvdA	(1, d, 1)	0.27*	−0.53*	0.44*	0.007	2,088.69	11.17
CDA	(0, d, 1)	—	−0.19*	0.35*	0.007	2,053.79	5.68

Note. ARIMA = autoregressive integrated moving average; BIC = Bayesian information criterion; LBQ = Ljung–Box Portmanteau; PvdA = Labor Party; CDA = Christian Democratic Appeal.

*$p < .05$.

nonsignificance is still slow in the detrended case (albeit not as slow as in the original series). Table 2.6 shows the best fitting models for both parties. An ARFIMA (1, d, 1) offers the best description for the PvdA results and an ARFIMA (0, d, 1) best fits the CDA data, with differencing parameters of $d = 0.44$ and $d = 0.35$, respectively, indicating a high degree of persistence in both cases.

Analyzing the Residuals

As is the case for many other statistical techniques, a crucial step is an analysis of the residuals, which will show whether our modeling efforts were successful or whether there are trends that remain unaccounted for. Note that in a time series analysis, the residuals of an ARIMA model are also a time series that can be subjected to further analysis. To confirm that models fit the data well, we expect the residual time series to display randomness of the kind shown in the top row of Figure 2.1, and an ACF plot showing randomness in the ACF as well.

Based on the script provided in the appendix to this chapter (available on the accompanying website for this book at **https://study.sagepub.com/ researchmethods/qass/koopmans-using-time-series**), the reader can generate the residual plots for each of the analyses conducted here and confirm that the models discussed here fit the data well. The residuals of the births to teens data are presented here, and they exemplify the value of conducting this type of analysis. The model comparisons shown in Table 2.5 suggest that both ARIMA (1, 0, 1) × (0, 0, 1)$_7$ (Model IV) and ARIMA (1, 0, 1) × (1, 0, 1)$_7$ (Model V) are said to fit the data well according to the BIC and

Figure 2.7   Residual Time Series and Autocorrelation Function (ACF) Plots of Two Models Characterizing the Births to Teens Data: Models IV and V (see table 2.5)

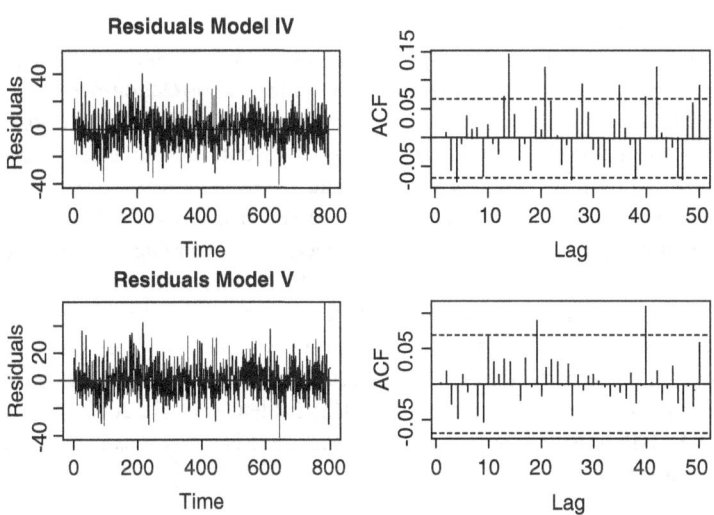

LBQ indicators. The residuals shown in Figure 2.7 qualify this conclusion. While the time series plots shown on the left panels of the figure look random enough for both models, the ACF plots in the right indicate that an autocorrelative pattern remains to be accounted for in Model IV, as can be seen by the spikes at Lags 14, 28, 35, and 42. The estimation of an additional seasonal autoregression parameter (Model V) remedies this situation, as can be seen in the ACF plot for that model, in which the two spikes that remain statistically significant seem incidental and are randomly distributed across the 50-lag spectrum shown in the plot. We therefore conclude that Model V is the best one to account for these data. Comparison of the BIC indicators in Table 2.5 already suggested as much.

Interpretation of the Differencing Parameter

While some scholars treat pink noise primarily as a statistical covariate (e.g., Clarke & Lebo, 2003), others suggest that it may indicate self-similarity in the system of interest, which, in turn, may point to an adaptive process between the system and its surroundings that forces it to deviate from regular behavioral cycles (Bak et al., 1987; Beran, 1994; Mandelbrot, 1997; Stadnitski, 2012a). As the results of the stationarity tests indicate for series containing pink noise, there are both stable and unstable aspects to those series, suggesting that while the system of interest is cohesive in the sense that the behavioral response is self-similar across cycles, it is also adaptive and self-organizing in the sense that the time scale over which these self-similar patterns are repeated fluctuates across iterations, depending, presumably, on the changing circumstances under which the adaptive process plays out. In the same way that finding a correlation between two variables may not be taken as expressing a causal relationship in the absence of any theory or other evidence, finding persistence or antipersistence in a time series by itself will not establish that the system is self-organizing or adaptive to changes. In other words, there is a risk of overinterpretation in these situations. At the very least, however, such a finding would indicate that the evidence does not permit us to rule these processes out. Plots such as those shown in Figure 1.3 are helpful in this context, as are the power spectra whose generation is further discussed in Chapter 3. Rigorous causal modeling of outcome time series to predictor time series addresses another aspect of this issue (e.g., Molenaar, 2015; Molenaar et al., 2009), but such models fall mostly outside of the scope of this book, save for a brief discussion in Chapter 5.

The Hurst Exponent

It is common parlance in dynamical circles to express metastability in time series in terms of the *Hurst exponent* rather than the differencing parameter. Hurst developed the parameter in the context of his hydrological research, which involved an estimation of river discharges of the River Nile to estimate seasonal variations in water flow and in the needs of local agri-culture, thus requiring an estimation of long-range dependencies in the time series (Feder, 1988; Hurst, 1956). Hurst's original computations involve an analytical technique called *Rescaled Range Analysis*, which is further dis-cussed in Chapter 4. Within the fractional differencing framework (which was developed much later), the Hurst exponent can be readily derived from the differencing parameter as follows:

$$H = d + 0.50. \qquad (2.12)$$

Since persistence or antipersistence in a time series is decided by testing whether $d \neq 0$ given that $-0.5 < d < +0.5$, it follows that $H = 0.5$ indicates the absence of persistence, while $0 < H < 1$ defines the continuum from antipersistence to persistence (Stadnitski, 2012a). In the nonstationary case, $H = 0.5$ if the increments in the differenced time series are uncorrelated; $H > 0.5$ if they are positively correlated; and $H < 0.5$ if the correlation is negative (Mandelbrot & van Ness, 1968). Obviously, the cautionary remarks made above concerning the interpretation of the differencing parameter apply equally to the interpretation of the Hurst exponent.

D. Chapter Summary and Reflection

As in any statistical analysis, the key to effective fractional differencing is the recognition of typical data patterns in the long range of the series to guide the modeling effort. To that end, I simulated six of such patterns in this chapter: (1) white noise, (2) short-range dependency, (3) pink noise, (4) Brownian motion (random walk), (5) a seasonal cycle at 7 lags, and (6) a seasonal cycle at 12 lags. This chapter described the extensions to the traditional regression-based time series analysis that permits us to distinguish these patterns. Four real data sets were then introduced to demonstrate the utility of this approach. There are three major challenges to the analysis of fractal patterns in time series data through fractional differencing. The first one is that the coefficients associated with long-range dependency and those associated with short-range dependencies are often heavily correlated and, thus, tend to mimic one another in the model diagnostic phase of the analysis. The stepwise modeling strategy described in this chapter is the best available solution to this problem at this point. It allows for an estimation of the impact of the fractal process over and above that of regular short-range and seasonal dependencies. The importance of such comparative modeling is illustrated by the finding that the analysis of the births to teens data set yields a statistically significant differencing parameter, but only if the estimates of short-range processes are not included in the model. Second, a well-known drawback of the types of analyses discussed in this chapter is the need for large numbers of data points to be able to estimate long-range patterns. This concern was also raised in Chapter 1. The data sets discussed in this chapter each consist of approximately 800 data points. Samples of this size are not always readily available, and as of this writing, the literature has not yet articulated a clear point of view of how fractal analysis can or should proceed if the number of time points is significantly smaller. For the time being, then, the directive is to collect more data that meet the high-definition requirements for the detection of fractality.

One terminological clarification is in order here. While fractals and fractional differencing have a shared etymological root (both refer to fractions), they refer to different aspects of the processes discussed here. *Fractional* refers specifically to the values that the fractional differencing parameter estimate can take on when a fractional differencing analysis is undertaken (here between -0.5 and $+0.5$), not to be confused with nonfractional integration, which assigns integer values to the differencing parameter, depending on the order of the difference ∇Y_t applied to the time series (see the section on integration). *Fractals*, on the other hand, are defined much more broadly in terms of self-similarity—that is, patterns that replicate within themselves. This distinction brings us to the third challenge encountered when conducting fractional differencing, which is the question to what extent finding long-range dependencies by obtaining a differencing parameter that is not equal to zero is a sufficient condition to conclude fractality. The literature does not offer a consistent point of view on this issue. Some investigators, like Mandelbrot (1997), are quite adamant that one can make this inference based on long-range dependencies. However, it will help support the conclusion of fractality to be able to triangulate the findings from fractional differencing with those from other approaches, such as PSDA, which is the focus of the next chapter.

3

POWER SPECTRAL DENSITY ANALYSIS

One of the drawbacks of the fractional differencing approach outlined in Chapter 2 is that it relies on the stationarity assumption for the estimation of fractality by generating a differencing parameter estimate. In cases where a time series is nonstationary, the differencing parameter is typically fixed to the number of lags over which the differencing is undertaken to produce a series of stationary increments, which is to say the d-th difference $W_t = \nabla^d Y_t$. Instead of using d to estimate fractality in the series, it is used in those cases to remove nonstationarity from the series. This approach restricts our ability to examine fractal patterns in volatile time series, and it creates a discontinuity in our analytical approach toward fractal patterns in the stationary and nonstationary cases. In addition, the integration removes authentic features from the data. A comparison of the time series of differenced and nondifferenced U.S. unemployment data in Figure 2.5 helps appreciate this point.

To achieve greater continuity between the stationary and the nonstationary cases, dynamical researchers sometimes prefer to use an alternative approach, *power spectral density analysis* (PSDA) to analyze fractal or metastable patterns. PSDA proposes a continuum from stationarity to nonstationarity by way of metastability and relies on a single analytical framework to estimate fractality, irrespective of stationarity assumptions. Comparing the plots of Figures 2.1, 2.3, and 2.4 suggests this kind of continuity as a gradual shift from randomness to the more clustery patterns that are seen in cases of short- and long-range dependencies. One of the main features of the PSDA approach is use of *power spectra* to assist with the estimation of fractal patterns in time series data. As with other analyses of fractal patterns, PSDA requires a large number of successive observations.

Power spectra were originally formulated to describe the inverse relationship between the frequency of occurrence of certain events and the magnitude of those events. According to this function, called the *power law*, declining frequency is correlated with increasing magnitude in a downward slope. For example, Bak (1996) described such a relationship between the frequency with which earthquakes occur and their magnitude on the Richter scale, which follows an almost perfect straight line from tiny quakes being common to big ones being rare. The larger implication of this work is that PSDA is a tool that can be used to model extreme events instead of assigning them outlier status such that they fall outside the purview of our models. The idea of correlating the frequency with which events replicate to the size of their impact on a system is also relevant to the understanding of time series with long-range patterns, and it is one of the central ideas behind PSDA. This chapter is structured as follows. I will start with a brief review of the basic approach and terminology of PSDA, and I will subsequently show how it is used in the analysis of sampled time series data, including the data presented in the previous chapter of this volume. I will then show how power spectra are generated and interpreted, and how they can be used to identify self-similar processes—that is, power laws, in the data of interest.

A. From the Time Domain to the Frequency Domain

Within the ARIMA analytical framework, a time series is analyzed as a set of sequentially ordered observations, in which outcomes at a given time point are regressed on previous observations (autoregression) and previous discrepancies from the mean of the series (moving average). The lag specificity of the modeling enables the investigator to estimate dependencies at lag sizes that may be substantively interesting, such as seasonal cycles. PSDA shifts the focus of our analytical efforts to the detection of cyclical patterns and the extent of their contribution to the overall variability in the series. These cycles express at which lags data points tend to be correlated to each other. Thus, the perspective changes from the analysis of the behavior of individual data points and their correlation to each other to the detection of cycles that repeat at varying frequencies. As will be shown below, sinusoid functions are then generated that best characterize the *frequency* at which these cycles *repeat* throughout the series. The extent to which cycles at given frequency levels characterize the variability in the series is called the *amplitude* or *power* of those cycles. The use of these two pieces of information, power and frequency, in conjunction is the defining feature of PSDA. We refer to this shift as a transformation from the time domain (time series) to the frequency domain (spectral density). An effective representation of PSDA benefits from a brief detour to the algebraic underpinnings of spectral analysis, which follows in the next section. This section can be skipped by the more globally oriented reader.

Amplitudes and Relative Frequencies

Spectral analyses are designed to model the rate at which a time series oscillates to describe the regularities in the fluctuation patterns. This oscillation is typically defined as a sine or cosine function as follows:

$$x_t = A\cos(2\pi\omega t) + \phi, \tag{3.1}$$

for $t = 0, \pm 1, \pm 2, \ldots$. In this equation, ω defines the number of cycles per unit, A determines the height or amplitude of the series and φ is the phase of the function (Shumway & Stoffer, 2011). This latter parameter is typically without a substantive interpretation. For purposes of data analysis, the following equation expresses this relationship more conveniently:

$$x_t = U_1 \cos(2\pi\omega t) + U_2 \sin(2\pi\omega t). \tag{3.2}$$

The variables and relationships defined in this equation are illustrated in Figure 3.1, based on a set of input values provided in Table 3.1 (Shumway & Stoffer, 2011, p. 177). The six panels in Figure 3.1 show the effect of varying the amplitude and relative frequency values on the appearance of the trajectories. The three trajectories on the left of the figure are labeled xa_1, xa_2, and xa_3. Those on the right are called xb_1, xb_2, and xb_3. On inspection of these panels, it can be verified that the relative frequency ω counts the number of cycles j relative to a given time frame (here, $n = 100$). Likewise, it can be seen in the figure that the amplitude A represents the deviation produced by these cycles. Thus, the size of the variation and frequency of the repetition are both captured in these functions, and they can be varied independently of each other. The squared amplitude and relative frequencies are indicated at the top of the panel for each function. The mathematical identities underlying the relationships shown in Figure 3.1 are summarized in the first section of Exhibit 3.1. This section shows the relationships between the components of Equation 3.2. These relationships can also be evaluated based on the values provided in Table 3.1.

It is possible for a time series to incorporate multiple cyclical patterns. To display such combinations of cycles, Equation 3.2 can be expanded to capture multiple frequencies and amplitudes as follows:

$$x_t = \sum_{k=1}^{q} U_{k1} \cos(2\pi\omega_k t) + U_{k2} \sin(2\pi\omega_k t). \tag{3.3}$$

This equation shows the prediction of x_t as a summation of cyclical components, such as those shown in Figure 3.1. Thus, a more complex series can be created that is made up of a set of $k = 1, 2, \ldots, q$ periodic components. Figure 3.1 shows such isolated periodic components. The two top panels in Figure 3.2 show the series that results from the sum of these components for $\sum_i xa = xa_1 + xa_2 + xa_3$ (*left*) and $\sum_i xb = xb_1 + xb_2 + xb_3$ (*right*).

Figure 3.1   Simulated Periodic Components With Varying Amplitude A and Relative Frequency ω

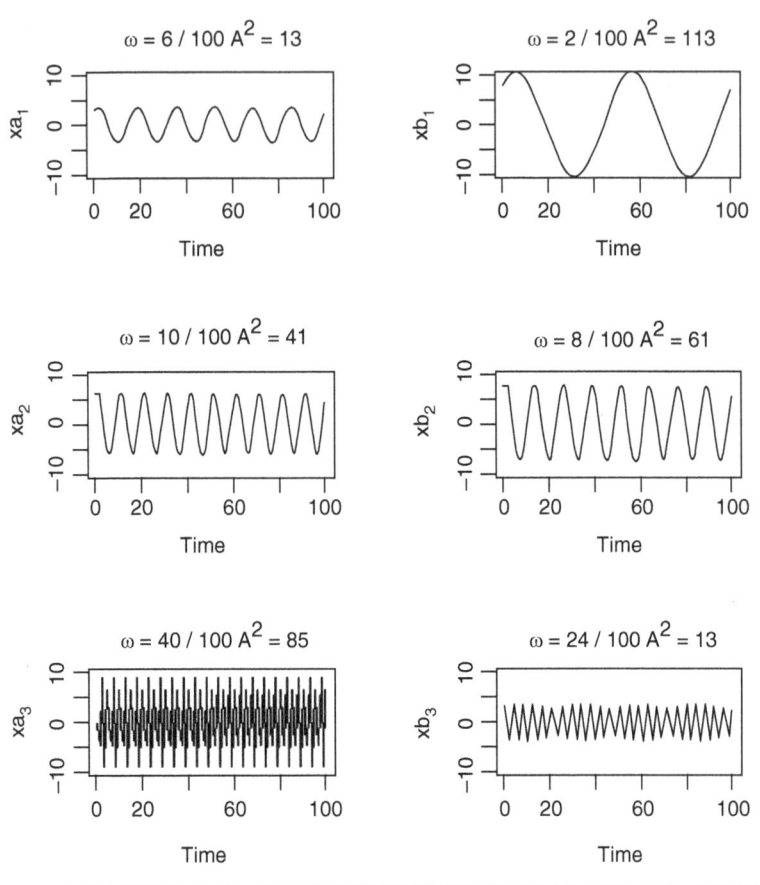

Source. Adapted from Schumway and Stoffer (2011, p. 177).

The Fourier Transform

The conversion from the time domain to the frequency domain involves a mathematical operation called the *discrete Fourier transform* (Mandelbrot & van Ness, 1968; Shumway & Stoffer, 2011), which reexpresses the trajectory of observations ordered on a time scale as a power versus frequency relationship in the following manner:

$$d\left(\omega_j\right) = n^{\frac{1}{2}} \sum_{t=1}^{n} x_t e^{-2\pi i \omega_j t}, \tag{3.4}$$

for $j = 1, 2, \ldots, (n-1)/2$, where $\omega_j = j/n$ are the relative frequencies. This equation shows the sum of the x_t over a total of n observations, as well

Table 3.1 Simulated Input Conditions for the Plots in Figure 3.1

Input	U_1	U_2	A^2	$\omega = j/n$	A	ϕ
Left Panels						
xa_1	2	3	13	6/100	3.61	−56.31
xa_2	4	5	41	10/100	6.40	−51.34
xa_3	6	7	85	40/100	9.22	−49.39
Right Panels						
xb_1	7	8	113	2/100	10.63	−48.81
xb_2	5	6	61	8/100	7.81	−50.19
xb_3	2	3	13	24/100	2.61	−56.31

Source. Adapted from Schumway & Stoffer (2011).

Exhibit 3.1 A Few Helpful Identities

1. In the periodic process shown in Equation 3.2, we define:

$$U_1 = A \cos\phi$$
$$U_2 = -A \sin\phi$$
$$A = \sqrt{U_1^2 + U_2^2} \quad \phi = \tan^{-1}\left(-\frac{U_2}{U_1}\right), \text{ provided that } U_1 > 0$$
$$\omega_j = j/n$$

2. In the discrete Fourier Transform shown in Equation 3.4, we define:

$$e^{i\lambda} = \cos(\lambda) + i \sin(\lambda)$$
$$\cos(\lambda) = \left(e^{i\lambda} + e^{-i\lambda}\right)/2$$
$$\sin(\lambda) = (e^{i\lambda} - e^{-i\lambda})/2i$$
$$\lambda = 2\pi\omega_j$$
$$i = \sqrt{-1}$$

Source. Adapted from Bloomfield (1976) and Schumway & Stoffer (2011).

48

Figure 3.2 Sum of the Periodic Components from Figure 3.1: Time Series and Scaled Periodograms

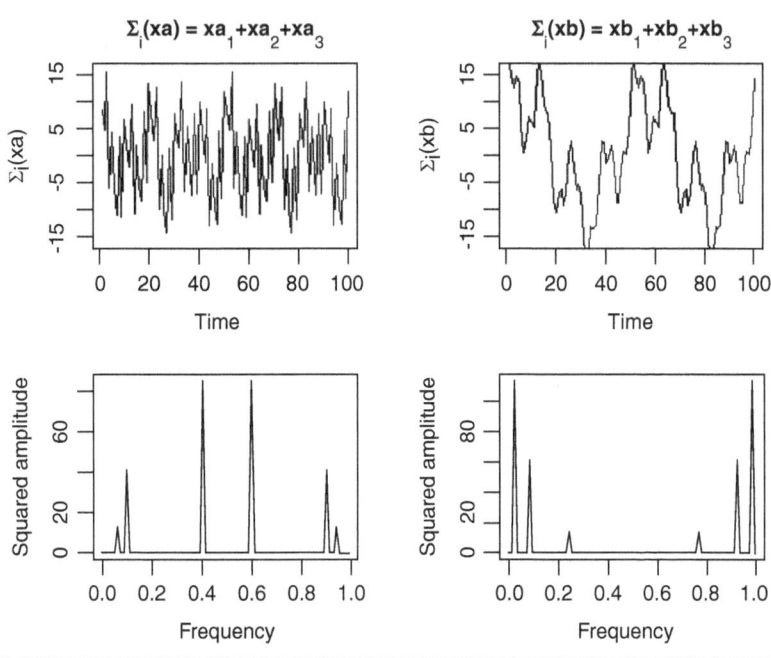

Source. Adapted from Schumway & Stoffer (2011).

Note. Series *xa* (*left*) and *xb* (*right*). Time series (*top row*); scaled periodograms (*bottom row*).

as the mathematical operations through which these values are transformed into the frequency domain. The second section of Exhibit 3.1 shows the mathematical identities that are relevant to this transform, on account of which we can also express Equation 3.4 as follows:

$$d\left(\omega_j\right) = n^{1/2}\left(\sum_{t=1}^{n} x_t \cos\left(2\pi t\omega\right) - i\sum_{t=1}^{n} x_t \sin(2\pi t\omega)\right). \quad (3.5)$$

The Fourier transform reconstructs a time series into cyclical components to estimate the amount of variance that is absorbed within the series by each of these components (Delignières et al., 2005; Eke et al., 2000; Stadnitski, 2012a; Stewart, 2012). As in the examples discussed above, the frequency *j* in Equation 3.1 refers to how often the cycles repeat; the relative frequency ω_j is the ratio of *j* to the total number of observations *n* in the series. Note that there cannot be more than *n/2* cycles, given that two data points are needed to produce a cycle. The cycle that repeats most frequently is the dependency between an observation and its immediate neighbor, which according to Equation 3.1 has a relative frequency of $\omega = \left(\dfrac{n-1}{2}\right)/n$, which approximates 0.5 as n gets larger. The cycle occurring least frequently is $\omega = \left(\dfrac{1}{2}\right)/n$, which approximates zero as n gets larger. For the most vivid explanation of the Fourier transform, how it works, and its place in human civilization, see Stewart (2012). Shumway and Stoffer (2011, p. 187) provide some computational exercises to do the transform in R.

Periodograms

The stated purpose of PSDA is to detect cyclical behavior in a time series and fit the sinusoid curve that best characterizes the cycles. Shumway and Stoffer (2011) define the periodogram as follows:

$$I\left(\omega_j\right) = \left|d\left(\omega_j\right)\right|^2, \qquad (3.6)$$

which is to say that the periodogram represents the absolute squared values of the Fourier-transformed series. A *scaled periodogram* is then computed as

$$P\left(\omega_j\right) = \frac{4}{n} I\left(\omega_j\right). \qquad (3.7)$$

The scaled periodogram plots the squared amplitude A^2 on the relative frequency ω, and thus, it is a useful diagnostic tool to detect underlying cycles in a time series. The bottom two panels of Figure 3.2 illustrate this point by showing how the squared amplitudes for xa and xb reproduce the cycles shown in Figure 3.1, on which they are based. For example, among the cycles on the left of the figure, xa_1 has a squared amplitude of $A^2 = 13$ and a relative frequency of $\omega = \dfrac{6}{100} = 0.06$. Likewise, xa_2 has a squared amplitude of $A^2 = 41$ and $\omega = \dfrac{10}{100} = 0.10$. The corresponding values for xa_3 are $A^2 = 84$ and $\omega = \dfrac{40}{100} = 0.40$. The scaled periodogram at the bottom left of Figure 3.2 shows these cyclical patterns as peaks of A^2 at the appropriate values of ω. The periodogram at the bottom right of the figure likewise illustrates for xb how the high squared amplitude of $A^2 = 113$ is observed at the low frequency of $\omega = \dfrac{2}{100}$, that is, xb_1, while lower amplitudes are seen for the higher frequencies at xba_2 and xb_3. Note that the sections of the periodogram below and above $\omega = 0.5$ mirror each other, and thus, it is sufficient to focus only on $0 < \omega \le 0.5$ in this figure.

Power Spectral Density

Power spectra are based on the spectral density. Spectral density analysis seeks to represent the distribution of covariance across the continuum of relative frequencies (Cryer & Chan, 2008). R computes it as

$$\hat{S}(f) = \frac{1}{2} I(\omega_j). \tag{3.8}$$

Power spectra are generated by taking the natural logarithm of the (non-scaled) periodogram value $S(f)$, and of the relative frequencies ω, and plot the former on the latter. With respect to power spectra, a more typical notation in the dynamical literature is to define the log periodogram versus the log relative frequency in the following relationship (see, e.g., Delignières et al., 2005):

$$S(f) \propto 1/f^\beta. \tag{3.9}$$

The superscript β in Equation 3.9 is also called the *power exponent*, which is the absolute value of the slope of the inverse relationship between power and frequency, as shown in the power plots.

Power spectra can assist us with the detection of long-range nonrandom patterns in the data. We say that there is a power law if the data in these spectra show a linear pattern with a downward slope. Figure 3.3 shows the nonscaled periodogram and the corresponding power spectra for four of the six simulations that were discussed in Chapter 2: white noise, short-range autoregression, pink noise, and Brownian motion simulated as a random walk. The randomness of the white noise process is discernible in both plots. The periodogram shows a random distribution of values across the frequency spectrum, while the power spectrum shows no particular upward or downward trend. The periodograms for the other three data scenarios show higher values at the relative frequencies toward the lower end of the continuum, although this relationship is much more pronounced in the cases of pink noise and Brownian motion, attesting to the importance of the long-range dependencies in those latter two cases. It can be verified, by the way, that the values on the abscissa of the power spectra are a log transformation of those on the abscissa of the periodograms immediately above in the figure. Likewise, it can be checked that if the periodogram values on the y-axis are divided by half on account of Equation 3.8 and log transformed, the values on the y-axis of the power spectra result.

We typically expect the power spectra to show an inverse linear relationship between the log power and the log relative frequency, and we expect this relationship to be stronger in the nonstationary case than in the case of

51

Figure 3.3 Periodograms and Power Spectra for Four Typical Data Scenarios: White Noise, Short-Range Autoregression, Pink Noise and Random Walk

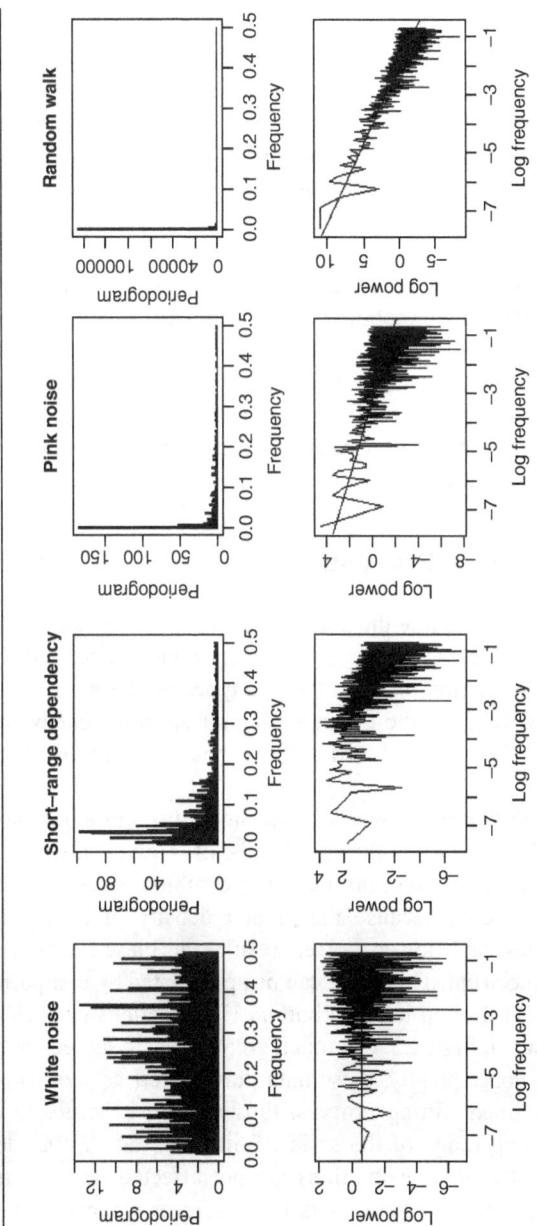

Note. The *top row* shows the periodograms; the *bottom row* shows power spectra with the following slope values: $\beta = 0.00$ (white noise), $\beta = -0.76$ (pink noise), $\beta = -1.87$ (random walk). No linear function is fitted for short range autoregression (see text).

pink noise. The power spectra for these latter two data scenarios confirm this expectation. The downward slope for the random walk is of a greater magnitude than that for pink noise ($\beta = -1.87$ and $\beta = -0.76$, respectively), and it can also be seen that the data points are clustered more tightly around the fitted line in the case of the random walk, indicating the stronger presence of long-range dependency in the series. It is important to note as well that there is no linear function to characterize the power for short-range autoregression. Instead, it appears that at the lower log frequencies, the relationship with amplitude is linear, while at higher log frequencies, it is not. This finding is consistent with the expectation of short-range but not long-range dependency (Stadnitski, 2012a; Wagenmakers et al., 2004). The pink noise and Brownian motion functions, on the other hand, show dependency over the long range of the series, and white noise function shows randomness throughout. To determine whether the slope of a power function is significantly different from zero, R conducts the hypothesis test within a conventional regression framework assuming a normal distribution of the parameter values.

B. Spectral Density in Real Data

In this section, I will generate the periodograms and power spectra for the four data sets that were also discussed in the previous chapter and provide a meaningful interpretation for what the two types of plot are showing. In addition, the consistency of the findings from the spectral density analysis with those from the factional differencing analysis will be discussed as well.

First of all, note that the power spectra shown in Figure 3.3 show the results for the all $n/2$ cycles in the simulated data. However, when generating power spectra based on real data, it is common practice to draw a sample from these data because the greater density of data points at higher frequencies and lower power results in biased estimates of the slope of the spectrum. This point can be appreciated by comparing the density of points at the top left and bottom right sections of each of the four spectra shown in Figure 3.1. A much greater density is seen at higher frequencies, and consequently, these data points exert an overwhelming influence on the linear fitting process, thus making it harder to detect cycles over the long range of the series. Wijnants et al. (2012) demonstrate that while the slope estimations are not affected when simulated data are used, the impact of this bias is significant when real data are used, resulting in an overrepresentation of data points for the short-range cycles. Given that long-range dependency is a primary concern when conducting these analyses, many researchers (e.g., Delignières et al., 2005, Stadnitski, 2012b) have responded to this dilemma by taking a fraction of the data at the lower frequencies—for example, the first 12.5% of the $n/2$ points. Wijnants et al. (2012) argue that this approach creates a dependency in the interpretation of power spectra on the sample size in the original series, and they argue instead for using a fixed sample size because it results in more stable estimates. In their examples, Wijnants et al. sample the 50 lowest frequencies; here the sample is taken at the lowest $2^6 = 64$ frequencies. It should be recognized, however, that in the end neither of these two solutions is really satisfactory, because they do not solve the underlying problem that the slopes of power spectra depend heavily on the size of the sample of data points taken irrespective of how it is determined.

The spectral density results for the daily count of births to teens, unemployment figures, attendance rates in School 1, and left–right political orientation among self-declared CDA and PvdA party affiliates are shown in Figure 3.4. The figure shows the periodograms in the top row and the power spectra in the bottom row. The births to teens periodogram on the left indicate significant short-range cyclical behavior at the relative frequencies that should correspond to a $1/7 = 0.14$ (weekly) and $1/3.5 = 0.29$, which marks half a week. Thus, the periodogram attests to a strong seasonal short-range cycle in these data. The power spectrum right below, also on the left, shows dependency at low frequencies, with the slope going down, but not at higher frequencies, which seem to be better captured by a straight line. The curvilinear pattern in the power spectrum supports the impression that short-range (high frequencies) and not long-range (low frequencies) dependencies are the prevailing pattern in these data. Therefore, the fitted slope value ($\beta = -0.71$) would be of limited substantive value, because it would suggest a long-range dependency that is not supported by the data.

The power spectrum for the unemployment figures shows a characteristic Brownian motion pattern, with the data points tightly clustered around the fitted line and a relatively steep downward slope ($\beta = -2.62$). The corresponding periodogram for these data shows the overwhelming significance of the long-range patterns over the short-range ones, a pattern that is also characteristic of Brownian motion and related nonstationary patterns. Based on the script provided in the appendix to this chapter (available on the accompanying website for this book at **https://study.sagepub.com/ researchmethods/qass/koopmans-using-time-series**), the reader can verify that these patterns will be much less pronounced in the power spectrum if the first difference is taken of the unemployment data. The attendance and left–right orientation data show a characteristics pink noise pattern in both

54

Figure 3.4 Periodograms and Power Spectra: Births to Teens, U.S. Unemployment, School Attendance and Detrended Left-Right Orientation (CDA and PvdA)

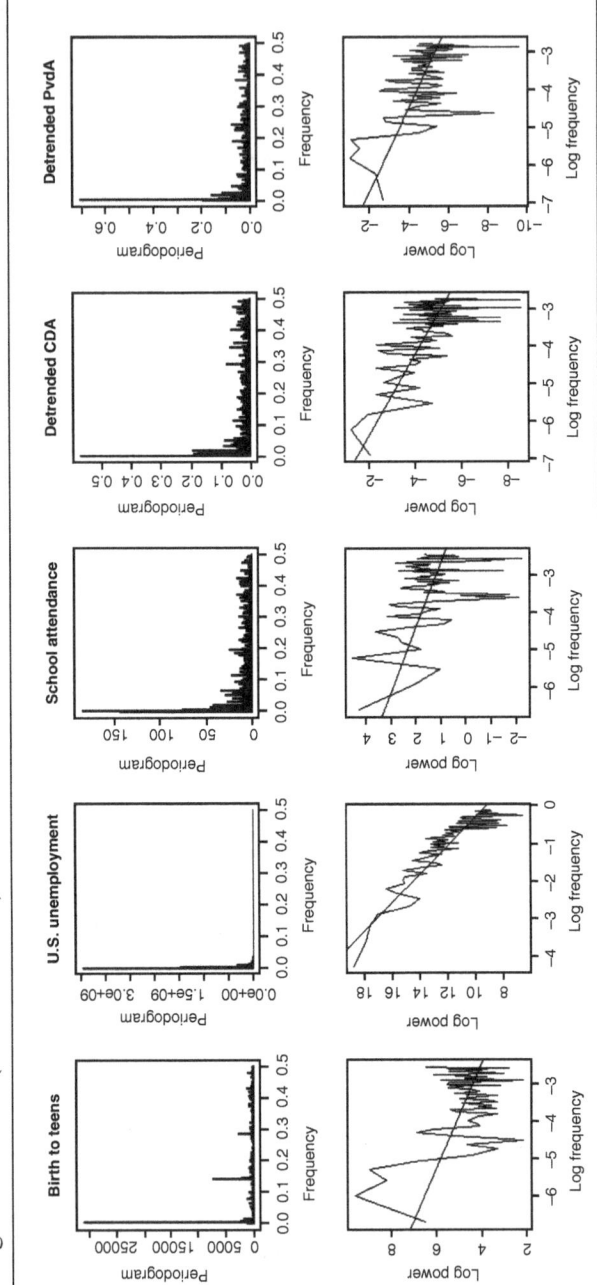

Note. Periodograms (*top row*); power spectra from the power spectra from the *bottom left* to *right* are: $\beta = -0.71$, $\beta = -2.62$, $\beta = -0.58$, $\beta = -0.93$ and $\beta = -0.90$, respectively. CDA = Christen-Democratisch Appèl (Christian Democratic Appeal); PvdA = Partij van de Arbeid (Labor Party).

plots. The slopes are well below zero in all three instances, indicating dependency throughout the relative frequency spectrum with a clear linear trend in the power spectra. The respective slopes in these three instances indicate the strength of these relationships: β = –0.58 (attendance), β = –0.93 (left–right orientation among CDA affiliates), and β = –0.90 (left–right orientation among PvdA affiliates).

It is important to appreciate the consistency of these results with those obtained through the fractional differencing approach described in the previous chapter, which reported persistence in the school attendance and left–right orientation data, a strong seasonal pattern in the births to teens data, and volatility in the unemployment data. It is instructive to take a closer look at the school attendance data. A tiny spike in the periodogram of the attendance data at the frequency of $1/5 = 0.2$ and the downward gap in the corresponding logarithm value in the power spectrum point to the possibility of a cycle according to the 5 days of the school week. The fractional differencing results shown in the previous chapter (Table 2.4) also point to this possibility. The power spectra for the attendance and political orientation data clearly show evidence of persistence (i.e., downward slopes, linear patterns). This result is consistent with those reported in Table 2.4 for attendance and Table 2.6 for political orientation. The teen pregnancy results, on the other hand, indicate in both types of analyses that the regular patterns override any signs of metastability that these data may contain, in spite of the fact that estimating the differencing parameter by itself would lead one to conclude persistence (see Table 2.2). This result underscores a more general issue with these types of analyses, which is that estimating irregularity without taking the regular patterns into account can be misleading (Wagenmakers et al., 2004). In Chapter 4, I will return to this point.

C. Fractional Estimates of Gaussian Noise and Brownian Motion

As mentioned previously, the advantage that PSDA has over fractional differencing is that it allows for the analysis of stationary and nonstationary series within a single analytical framework, without having to resort to stationary increments when data are volatile. PSDA nonetheless makes a distinction between the stationary and nonstationary case in the analysis of Fourier-transformed data to detect fractal patterns. Within this analytical framework, the underlying pattern in these two cases is referred to as fractional Gaussian noise (fGn) and fractional Brownian motion (fBm), respectively (Eke et al., 2000). However, while the determination of stationarity in time series data is based on the ADF and KPSS tests discussed in Chapter 2 (see Table 2.1), the distinction between fGn and fBm is based here on the slope of the power spectrum in the following manner: $+1 < \beta < -1$ (fGn) and $-1.00 < \beta < -3.00$ (fBm). As is the case with fractional differencing, the parameter favored in the dynamical literature for the detection and interpretation of fractal patterns is the Hurst exponent H. There is a variety of ways in which the Hurst exponent can be generated from time series data (see Delignières et al., 2005). In Chapter 2, it was shown how H can be derived from the fractional differencing estimations; PSDA expresses H in terms of the slope of the power function. In this instance, the slope of the power function is the inverse of the power coefficient in Equation 3.9. Contrary to fractional differencing, which requires the stationarity assumption for the estimation of H, PSDA permits the estimation of H in both stationary and nonstationary series. However, its computation differs depending on whether the series is designated as fGn or fBm in the following manner:

For fGn,

$$H = (|\beta| + 1)/2, \tag{3.10a}$$

and for fBm,

$$H = (|\beta| - 1)/2. \tag{3.10b}$$

Based on the slope values for the power spectra shown in Figure 3.4, Table 3.2 compares how H is estimated based on fractional differencing and based on the results of the PSDA for the attendance data, unemployment data, and political orientation data for the CDA and PvdA affiliates. Due to the curvilinear pattern in the births to teens data, no Hurst exponent was generated in that case because the linear power function does not adequately characterize these data. In the unemployment data, the undifferenced trajectory is used, because it enables the power function to capture the volatility in the data. However, fractional differencing assumes stationarity, and therefore, the Hurst exponent cannot be estimated based on the original data, and it is therefore not included in the table. It was discussed in Chapter 2 how taking the first difference produces a stationary series, whose short-range autoregressive and moving average parameters can be estimated. However, the estimation of H on the basis of that information would create a result that is not compatible with the findings of the power spectral analysis, because the input series look vastly different, as can be seen by comparing the unemployment trajectories in the second and third row of Figure 2.5.

Table 3.2 Estimates of the Hurst Exponent Based on Fractional Differencing and Power Spectral Density Results in the Real Data

Data Source	Fractional Differencing Estimates		Power Spectral Density Estimates	
	d	$H = d + 0.5$	β	$H = (|\beta| \pm 1)/2$
School 1 attendance	0.27	0.77	−0.58	0.79
Births to teens	0.26	0.76	—	—
Unemployment	1.00	—	−2.62	0.81
CDA (detrended)	0.35	0.85	−0.93	0.97
PvdA (detrended)	0.44	0.94	−0.90	0.95

The results in Table 3.2 indicate that for the daily attendance data in School 1, the fractional differencing and power spectral density results are quite compatible, with $H = 0.77$ and $H = 0.79$, respectively. Likewise, the Hurst exponents of the left–right orientation of the PvdA affiliates comes quite close for both approaches: $H = 0.94$ based on fractional differencing, and $H = 0.95$ based on PSDA, showing a degree of persistence in both cases that is quite close to nonstationarity. The case of left–right orientation among CDA affiliates shows a discrepancy in the results from the two estimation methods. Based on fractional differencing, the Hurst exponent comes to $H = 0.85$; based on power spectral density, $H = 0.97$. In both instances, then, the findings point to a strong persistence in these data, while a more than 10% discrepancy remains, which is quite substantial on a scale from 0 to 1. Note that in the case of the left–right orientation data, the detrended data are used. It is possible to fit our models to the original data. However, this approach would yield indicators of high volatility in these data that is largely attributable to the upward and downward trend in the data, rather than to any possible indicators of fractality in these data, and is therefore of limited use in the present context.

D. Chapter Summary and Reflection

Two distinct approaches to the analysis of fractal patterns in time series analysis are those that rely on the time domain and those that rely on the frequency domain. In the time domain, a time series is represented as a string of sequentially ordered observations, while in the frequency domain, those series get reconstituted such that the frequency at which cycles repeat gets correlated to the variance explained by those cycles, that is, the amplitude or power. The PSDA approach discussed in this chapter relies on the data as they reside within the frequency domain, in which a log transformation of the power and of the relative frequency yields a power spectrum. These spectra are of particular interest to dynamical scholars as the linear relationship between log frequency and log power shows evidence of fractality, which is to say that dependencies are evenly distributed across the series. This pattern, in turn, points to scale invariance, which is a defining characteristic of fractality (i.e., the patterns are the same while the scale is different). One major difficulty when conducting PSDA is that it is extremely difficult to distinguish short-range dependence and seasonal dependence from long-range dependence, and thus, the approach is not very suitable to estimate the importance of these variance components relative to one another. However, the analysis derives its value from the fact that power spectra actually show fractality in the data, where the inference of such from fractional differencing results is not straightforward. Thus, the case should be made for the use of fractional differencing and PSDA in conjunction such that they compensate for each other's shortcomings and complement their respective strengths.

Conventional time series makes a distinction between stationarity and nonstationarity depending on whether the statistical properties are constant across a series. Fractional differencing of the data requires that we can assume stationarity and otherwise, the time series gets converted to a string of stationary increments, typically but not necessarily at a lag size of one. In the frequency domain, on the other hand, a distinction is made between the underlying patterns of fGn and fBm, a distinction that roughly corresponds to stationarity versus nonstationarity, although the status of the data as being indicative of fGn or fBm is made on the basis of the slope of the power spectrum, input, in other words, that resides in the frequency domain. In light of the gradual nature of the differences between white noise, pink noise, and Brownian motion patterns, as shown in Figures 2.1, 2.3, and 2.4, there is an advantage to having the analyses for these two types of data reside within a single analytical framework such that the differences between pink noise and Brownian motion–like patterns are actually based on a comparison of the coefficients describing the data. It can be readily appreciated in those instances that the pink noise and Brownian motion patterns are really a stronger and a weaker expression of the same phenomenon shown, respectively, by a steep or less steep slope in the power spectrum. Thus, the attendance data, unemployment data, and political orientation data are readily comparable in terms of the strength of the relationship between power and frequency and, thus, in terms of their fractality.

4

RELATED METHODS IN THE TIME AND FREQUENCY DOMAINS

In the previous two chapters, the discussion focused on the detection of fractal patterns in time series data, either through lag-based regression models (Chapter 2) or by modeling cycles in the frequency domain to determine their scale invariance (Chapter 3). Several alternative approaches have been suggested in the literature, which are the focus of this chapter. These approaches fall into roughly two groups: (1) methods that estimate fractal variance in the time domain and (2) periodogram-based spectral regression methods that fall in the frequency domain. Comprehensive overviews of these two sets of approaches are provided by Taqqu et al. (1995), Delignières et al. (2005), Stadnytska et al. (2010), and Stadnitski (2012b). Examples of such analyses within the time domain are detrended fluctuation analysis (DFA; Peng et al., 1994), the rescaled range (R/S) analysis (Hurst, 1965), and Higuchi's (1988) estimation of the fractal dimension. Examples of analyses within the frequency domain are Geweke and Porter-Hudak's (1983) log periodogram regression, smoothed log periodogram regression (Reisen, 1994), and the Whittle estimator (Taqqu & Teverovsky, 1997). While these analytical approaches are not used on a wide scale and much of the work being done is still in the simulation stage, there are some significant applications to real data, such as Peng et al.'s (1994) use of DFA to analyze patterns of irregular heartbeat and Delignières et al.'s (2004) use of fractal variance methods to analyze self-esteem. Higuchi (1988) discusses the applicability of those methods to analyze interplanetary magnetic fields. This chapter provides a synopsis of each of these approaches and shows some applications using the unemployment data and political affiliation data discussed in previous chapters.

A. Estimating Fractal Variance

One possible way in which data can exhibit fractal patterns is that the variability in the data exhibits scale invariance, which is a linear dependency of the degree of variability in a time series on the length of the series. The analysis of such dependency requires that a given time series gets carved up in a set of overlapping windows of ever-increasing size, allowing for a comparison of the dispersion within each of these windows. It is argued that it is suggestive of fractality if the variance in the data increases in a linear fashion as the size of the frame within which this variance is computed gets larger. It is argued that findings of a linear relationship between variance and the size of the frame is a manifestation of nonrandomness pointing to scale invariance in the data (e.g., Delignières et al., 2005). Therefore, it is of interest to dynamical researchers as one of the underlying patterns of fractality in time series. In this section, several variations of this approach are discussed. *Detrended fluctuation analysis (DFA)* computes the root mean square error in intervals of ever-increasing size to determine if the relationship between these two variables displays a linear pattern. The *rescaled range (R/S) analysis*, similarly, estimates this relationship using the range within each interval. *Higuchi's fractal dimension*, finally, extracts from a given time series a set of new trajectories based on different lags between the individual data points in the original series in order to decide whether the sum of absolute differences between the adjacent pairs in the new trajectories are proportional to the length of those trajectories. In each of these instances, the data points in the time domain form the basis for the determination of fractality in the data. Below is a more detailed description of each method.

Detrended Fluctuation Analysis

DFA partitions a time series into time frames, or blocks, of ever-increasing size. This process generates a set of sampled series with varying numbers of observations. Within each block or time frame, the series gets detrended. DFA then concerns itself with the question of whether a power law relationship is seen between the error variances within each block as the size of the blocks is multiplied. Finding such a relationship is an indicator of scale invariance, but it conceptualizes this invariance in terms of the cumulated discrepancy from the (detrended) mean within each block, rather than the actual measurements or their amplitude.

DFA relies on the cumulative sum of each observation x_t from the mean of the series X_t as follows:

$$X_t = \sum_i^T (x_t - \bar{x}). \tag{4.1}$$

The series generated by Equation 4.1 is then carved up in nonoverlapping blocks of size n. Within each of these blocks, a regression model is fitted and the residual variance F around the fitted values is calculated as follows:

$$F = \sqrt{\frac{1}{n}\sum_{t=1}^n (X_1 - X_n)}. \tag{4.2}$$

F is expected to increase proportionally with the block size n according to the following power law:

$$F \propto n^\alpha. \qquad (4.3)$$

In this function, α estimates the slope of the plot of the logarithm of F on the logarithm of the scale whose variability is represented (i.e., the size of the sampling blocks, see Delignières et al., 2005). If a straight line can be fitted through this plot with a nonzero slope value, this result is said to indicate scale invariance, or self-similarity in the variance patterns. DFA differentiates fGn and fBm, where fGn corresponds to power exponents ranging from $0 < \alpha < 1$ and fBm corresponds to $1 < \alpha < 2$.

Figure 4.1 shows an example of the DFA results for three of the simulated data scenarios also discussed in previous chapters: (1) white noise, (2) pink noise, and (3) the random walk. The cumulative sum is taken here for each of the series as shown in Equation 4.1. The resulting new series is shown in the top three panels of the figure. Peng et al. (1994) present DFA as an attempt to distinguish real fractality in the data from random processes that mimic fractality. It is important to realize in this context that fGn and fBm are reversible processes: taking the nth difference of fBm generates fGn, while the cumulative sum of fGn creates fBm (Delignières et al., 2005). Figure 2.3 shows how fGn can be created by taking the first difference of an fBm process. The top left panel of Figure 4.1 shows how the first-order cumulative sum of a random fGn process results in fBm. In light of this finding, the search for authentic fractal processes involves a comparison between the cumulative sum of a random process and processes of long-range dependence, such as pink noise and Brownian motion.

The panels at the bottom of the figure illustrate the power law relationship between the log residual variance and the log scale size. There is a linear relationship in all three instances. Note that the investigator can control size of the scales used to fit the power function. Here, we rely on the default settings in R, which is to say that for a series of the length N, the scale size ranges from 4 through $N/2$ at a ratio of 2, that is, 4, 8, . . ., $N/2$, or the natural log thereof, as shown in plots at the bottom row of Figure 4.1. When reporting findings for DFA analyses, it is important to specify these input settings.

Comparing the estimates produced by DFA, as shown in Figure 4.1, helps appreciate the difference between the cumulative sum of a random pattern and processes with genuine long-range dependencies. For fGn, the values of α range from 0 to 1, while for fBm, α range from 1 to 2. In the simulations shown here, the estimates are, respectively, $\alpha = 0.55, 0.98$, and

62

Figure 4.1 Detrended Fluctuation Analysis of Simulated Data: White Noise, Pink Noise and Random Walk

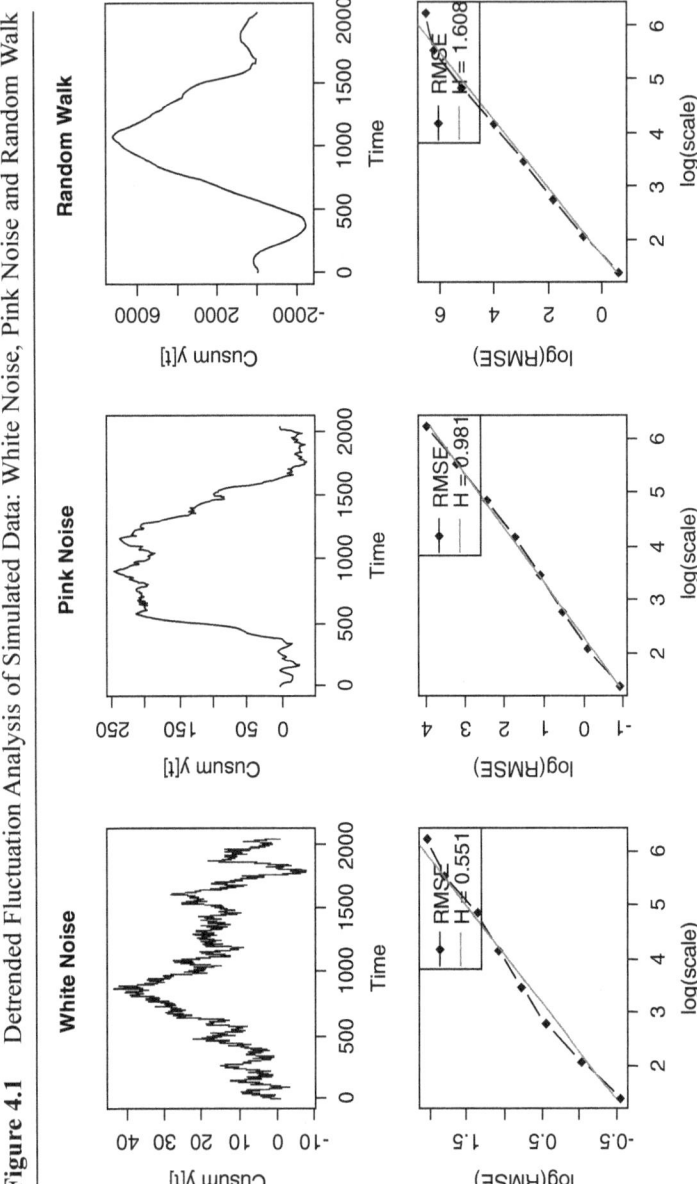

Note. Cumulative sum (Cusum) of the time series Y_t (*top row*); Extended Data Analysis (EDA) plots (*bottom row*). In this output, H represents the slope or power exponent of the relationship, not the Hurst exponent (see text). RMSE = Root mean square of error

1.61, for white noise, pink noise, and the random walk, respectively. The larger values of the alpha in this example indicate a higher degree of scale invariance in the detrended variability in the two nonrandom processes simulated here, an indicator of a fractal process. The Hurst exponent can be generated from these estimates as follows. To obtain the Hurst exponent from this analysis for fGn, the estimation is

$$H = \alpha; \tag{4.4}$$

for fBm, it is

$$H = \alpha - 1, \tag{4.5}$$

see Delignières et al. (2005). Thus, for the three simulations shown here, $H = 0.55$ (white noise), 0.98 (pink noise), and 0.61 (random walk).

Figure 4.2 shows the DFA results for the U.S. unemployment numbers and for the left–right political orientation findings for survey respondents identifying as affiliates of the CDA and PvdA political parties. These data sets were also discussed before. The top panels of Figure 4.2 show the cumulative sum of the time series for each of these three data installments. It can be seen that the volatility largely disappears from the unemployment data, while the detrended CDA and PvdA data show a high degree of volatility once the first cumulative sum is taken. This result was to be expected due to the stationarity of the detrended political orientation data. It can be seen that the alpha exponents reported here largely agree with the results obtained through the other analyses. The power exponents α for the left–right orientation data equal to 0.891 for detrended CDA and 0.944 for detrended PvdA. These values are close to 1, which is consistent with the metastability of the series reported in the previous chapters. Likewise, a power exponent for the unemployment data ($\alpha = 1.732$) indicates a high degree of volatility, also consistent with the findings reported previously.

Modifications have been proposed for DFA to enhance its reliability by Yazawa and Omata (2019). However, their software is proprietary and not for sale, and therefore, no applications are included in this text.

Rescaled Range Analysis

The R/S analysis was introduced by Hurst (1965), and it generates the Hurst exponent (H) in its original definition. Delignières et al. (2005) describe the method as follows. R/S divides a series X_t of the length N into

64

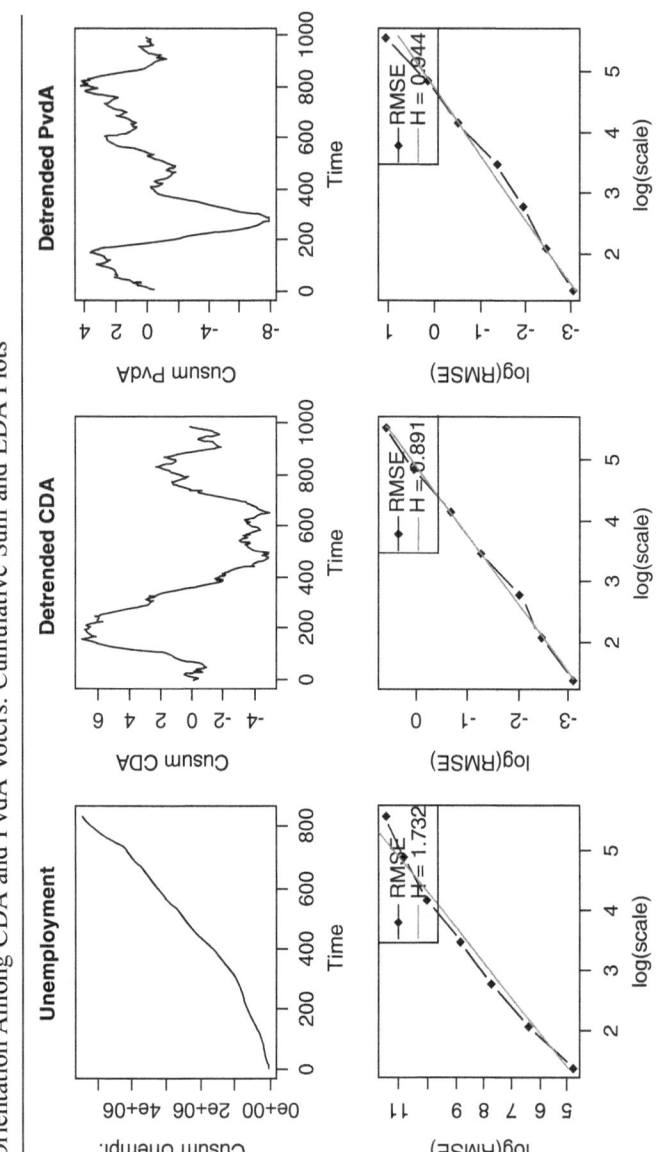

Figure 4.2 Detrended Fluctuation Analysis Results for Unemployment and Left–Right Political Orientation Among CDA and PvdA Voters: Cumulative Sum and EDA Plots

Note. RMSE = root mean square error; CDA = Christen-Democratisch Appèl (Christian Democratic Appeal); PvdA = Partij van de Arbeid (Labor Party); Unempl. = unemployment; Cusum = cumulative sum; EDA = extended data analysis.

nonoverlapping intervals of length n, and within each interval, an integrated series X_{tn} is computed:

$$X_{tn} = \sum_{i=1}^{t}\left(x_i - \bar{x}\right).$$ (4.6)

The range R is then defined as the difference between the *highest* and the *lowest* value of X_{tn} within each interval. The standard deviation is computed for the entire series, and R/S divides the range within each block by this standard deviation. These computations are repeated over a wide range of possible interval sizes, typically ranging from $n = 10$ to $n = (N - 1)/2$. The range, thus rescaled, is related to the size of the intervals by a power law relationship:

$$\overline{R/S} \propto n^{H}.$$ (4.7)

The power exponent in this equation is the Hurst exponent.

A statistical correction has been proposed to this method by Caccia et al. (1997) such that rather than calculating the discrepancies in Equation 4.6 based on the straight lines of the fitted average within each block, a detrended line is fitted that connects the end points of each interval. The discrepancies to that line are then computed instead. It should be noted that irrespective of the approach taken, R/S assumes stationarity.

Higuchi's Fractal Dimension

Feder (1988) notes that fractals look the same whatever the scale. We can look at the clouds in the sky or the broccoli on our plate to appreciate this point. The fractal dimension expresses the relationship between the size of the scales used to describe a given pattern and the number of scales needed to capture the pattern in its entirety. Taking the coastline of southern Norway as an example, Feder shows a linear relationship between the log size of the scales or boxes (δ) in kilometers and the log count of number of boxes $N(\delta)$ that is needed to cover the coastline in his example. This log–log plot is a fit of

$$N(\delta) = a\delta^{-D}.$$ (4.8)

In this equation, the term D defines the slope of this relationship, which is known as the *fractal dimension* (Feder, 1988). A more detailed account of the fractal dimension and its computation can be found in Kaplan and Glass (1995), whose discussion includes (biological) applications as well as challenging exercises and research project suggestions.

Mandelbrot (1997) extends the notion of fractality to time series data and its ability to describe self-similar patterns, such as those shown in Figure 1.3, and estimate financial risk. Since self-similarity in univariate time series is manifest over one coordinate dimension, not two or three, Mandelbrot prefers to use the term *self-affinity* in this context, a practice that has not been widely followed in the dynamical literature.

Higuchi (1988) introduces the fractal dimension of a time series as an alternative to the power law models, discussed in Chapter 3, arguing that irregular patterns in time series resist being captured by a single parameter, such as the slope of a power law, because these irregularities could be local. Therefore, an alternative is needed that captures this irregularity at scales of varying sizes. Higuchi's method carves up a trajectory X_N into a set of k new trajectories X_k^m. Within those, m denotes the point of onset for a given interval and k denotes the size of the intervals at which observations get picked up from the original time series for each new series.

Consider the series

$$x_1, x_2, x_3, \ldots, x_N . \tag{4.9}$$

A new series can be constructed as indicated below:

$$x_k^m = x_m, x_{m+k}, x_{m+2k}, \ldots, x_{m+\left(\frac{N-m}{k}\right)*k} . \tag{4.10}$$

Here, m is the time of onset and k is the size of the interval. So for example, if $k = 3$ and $N = 100$, a set of three new time series is sampled as follows:

$$x_3^1 = x_1, x_4, x_7, \ldots, x_{100}, \tag{4.11a}$$

$$x_3^2 = x_2, x_5, x_8, \ldots, x_{98}, \tag{4.11b}$$

$$x_3^3 = x_3, x_6, x_9, \ldots, x_{99} . \tag{4.11c}$$

The fractality of a time series is based on the *curve length* of the subsets, which is defined as follows:

$$L_m(k) = \left\{ \left(\sum_{i=1}^{N-m/k} \left| X(m+ik) - X(m+(i-1)k) \right| \right) \frac{N-1}{N-m/k} \right\} / k . \tag{4.12}$$

This equation defines the curve length in terms of the sum of the absolute differences between pairs of observations that are adjacent at the interval k given the onset point m of the series in the set. $L(k)$ represents the average curve length across trajectories $L_m(k)$. We say that the curve length is fractal if the following relationship holds:

$$L(k) \propto k^{-D}.$$ (4.13)

In R, the computation of Higuchi's fractality dimension requires the estimation of the Hurst exponent. Taqqu et al. (1995) show that the fractality dimension can then be recovered as follows:

$$D = 2 - H.$$ (4.14)

Figure 4.3 shows the fractal relationship advanced in Equation 4.8 between curve length and interval size in log–log plots for the simulated white noise, pink noise, and random walk data. The fractal dimensions for these three scenarios are, respectively, $D = 1.6$, $D = 1.2$, and $D = 1.1$. As before, a Hurst exponent of $H = 0.5$ indicates the absence of persistence, which roughly corresponds to the fractal dimension for white noise, while larger values of H in the other two cases yield correspondingly smaller values for the fractal dimension. The Hurst exponents on which the computations of D are based for the three simulation scenarios are, in the same order as above, 0.41, 0.80, and 0.87. This analysis tells us, in other words, that fractality can be found in varying degrees in these three scenarios but is pronounced in the cases of pink noise and Brownian motion, as was to be expected.

Figure 4.4 shows the fractal dimension in the political left–right orientation among CDA and PvdA voters in the Netherlands. These dimensions are 1.36 and 1.44, respectively, with corresponding Hurst exponents of 0.64 and 0.56. As reported previously, there is persistence in these data, but with much lower estimates than those reported based on fractional differencing and PSDA (see Table 3.2). The slightly nonlinear trend shown in the figure should inspire caution concluding fractality here, as it indicates that the fractal relationship as stated in Equation 4.8 may not hold here.

Table 4.1 shows the Hurst exponents generated by DFA, R/S, and Higuchi's method where appropriate for white noise, pink noise, and the random walk. There is a fair amount of variability across techniques in these results. In the case of white noise, R/S identifies some persistence ($H = 0.64$), while DFA detects barely any ($H = 0.55$), and Higuchi's fractal dimension even suggests a mild case of antipersistence ($H = 0.41$). The models concur on persistence in the case of pink noise, but do not agree on the strength of it, ranging from 0.77 (R/S) to 0.98 (DFA). The nonstationary results clearly indicate that different interpretations of H are required depending on the technique that is used to generate it.

68

Figure 4.3   Higuchi's Fractal Dimension in Simulated Data:
Logarithmic Plot of the Fractal Relationship between Curve Length and
Interval Size

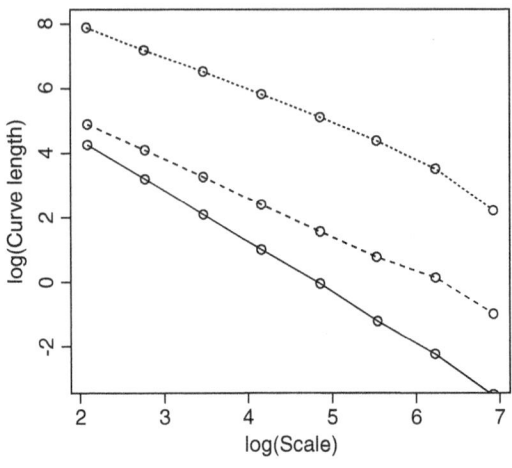

Note. White Noise (*D* = 1.6, *solid line*); Pink Noise (*D* = 1.2, *dashed line*); Brownian Motion
(*D* = 1.1, *dotted line*).

Figure 4.4   Fractal Relationship between Curve Length and Interval Size
in Left-Right Political Orientation in the Netherlands

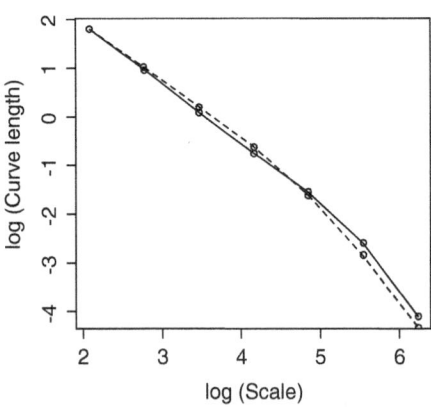

Note. The fractal dimensions are *D* = 1.36 for CDA voters (*solid line*) and *D* = 1.44 for PvdA
voters (*dotted line*). CDA = Christen-Democatisch Appèl (Christian Democratic Appeal);
PvdA = Partij van de Arbeid (Labor Party).

Table 4.1 Hurst Exponents (*H*) in Simulated White Noise, Pink Noise, and Random Walk: Results From DFA, R/S, and Higuchi's Fractal Dimension

Scenario	DFA	R/S	Higuchi's Fractal Dimension
White noise	0.55	0.64	0.41
Pink noise	0.98	0.77	0.80
Random walk	0.61	—	0.87

Note. DFA = detrended fluctuation analysis; R/S = rescaled range.

Related Approaches

Several variations have been proposed to the estimation of fractal variance, such as the use of the *aggregated variance* across blocks, the use of the *first difference of the variances* across blocks, or the use of the *absolute values of the variance* within blocks. Taqqu et al. (1995) provide an overview of these techniques. About these data, they tell essentially the same story and will not be further discussed here.

B. Spectral Regression

Fractality in time series data can also be estimated in the frequency domain through the use of periodogram-based regression models. The literature provides several examples of this approach, some of which generate a differencing parameter, while others compute a Hurst exponent. As is the case with the PSDA discussed in the previous chapter, the advantage of using these approaches is that they do not rely on the stationarity assumption and can capture fractality in both fGn and fBm, even if the differencing parameter is used. A general formulation of these models is provided by Beran (1994), who defines the shape of the spectral density function (see Chapter 3) as follows:

$$\log I\left(\omega_j\right) \approx \log c_f + \left(1 - 2H\right)\log \omega_j + \log \xi_j, \qquad (4.15)$$

where c_f is a positive constant and ξ_j are independent, normally distributed random variables. The other terms of Equation 4.15 are defined as in Chapter 3. Based on this relationship, a regression equation can be constructed as follows:

$$y_j = \log I\left(\omega_j\right), \qquad (4.16a)$$

$$x_j = \log \omega_j, \tag{4.16b}$$

$$\beta_0 = \log c_f - C, \tag{4.16c}$$

$$\beta_1 = 1 - 2H, \tag{4.16d}$$

$$e_j = \log \xi_j + C. \tag{4.16e}$$

Thus,

$$y_j = \beta_0 + \beta_1 x_j + e_j. \tag{4.17}$$

In this set of equations, C is the Euler constant, a quantity equal to 0.577215 as traditionally derived from number theory. The Hurst exponent can then be estimated as

$$\hat{H} = \left(1 - \hat{\beta}_1\right) / 2. \tag{4.18}$$

Geweke and Porter-Hudak (1983) formulate a periodogram-based regression model similar to Equation 4.15, but estimating d instead of H:

$$\ln I\left(\omega_j\right) = \ln\left(\frac{\sigma^2}{2\pi}\right) - d \ln 2(1 - \cos\left(\pi t \omega_j\right)). \tag{4.19}$$

Several varieties of spectral regression have been proposed in the literature, all of which are aimed at improving the modeling, which, in the original formulation provided above, tends to overrely on data at the higher end of the frequency range of the spectrum. You may recall that this issue surfaced as well in the context of PSDA, which fits a conventional linear regression function on the log–log power spectrum (see Chapter 3), and needs to correct for an overrepresentation of data points at the bottom right of the spectral plot. Here, as with power spectra, the issue can be addressed by drawing a sample of points for the power spectrum from the lower frequency range. Geweke and Porter-Hudak (1983) conduct regressions on the periodogram data, using a sample of 50 data points, which yield an estimate of the differencing parameter d, and is called the GPH estimator. Others have responded to the overrepresentation in the higher frequencies of the periodogram by partitioning the relative frequencies of the spectrum into blocks and compute the average spectral density within each of these blocks (smoothing). The results of this adjustment can then be used as a basis for model fitting (Reisen, 1994; Taqqu et al., 1995). In a related approach, Whittle's δ, estimates long-range patterns, based on a maximum-likelihood estimation, including confidence intervals (Beran, 1994;

Taqqu et al., 1995). Further details about the mathematical distinction of these methods is beyond the reach of this text and can be found in Taqqu et al. (1995), as well as in the original papers cited here.

A comparative study on the reliability of these types of estimates has been conducted by Stadnitski (2012a), who concludes that the fractional differencing approach, as discussed in Chapter 2, is decidedly superior to periodogram-based regression models, particularly since the latter overestimate fractality in the presence of short-range estimators, including seasonal ones. Table 4.2 illustrates this result in a set of simulated data that include fractional, seasonal, and random-only (white noise) variance components. In this table, the modeling strategies are compared: estimates of H using a smoothed periodogram and estimates of d using the GPH estimator (Geweke & Porter-Hudak, 1983), Sperio's estimate (Reisen, 1994), and Whittle's estimate (Beran, 1994). For purposes of comparison, DFA and fractional differencing results with and without short-range covariates are included in the table as well. Simulations were generated using the arfima simulation package in R, which can generate models that distinguish parameters associated with the estimation of long-range, short-range, and seasonal dependencies.

Table 4.2 shows the estimates of H or d obtained through each of these methods, as well as the expected values for these estimates under the specifications of the simulation models. These expectations are based on the input values for arfima, which is a fractional differencing method and the corresponding $H = d + 0.50$ (see Equation 2.12). It can be seen in the table that the generated estimates stay close to their expected values in the white noise scenario—that is, ARFIMA (0, 0, 0). The three simulations in which the value of H is specified, but not any short-range parameters, there are discrepancies both for H and d in the fitted DFA and smoothed spectral regression models. In case of the latter approach, none are very large, although a difference of 9 percentage points for the input of $H = 0.65$ is still quite substantial. DFA performs better at this smaller value of H but has greater discrepancies at higher input values of H. In the estimation of the Hurst exponent, the differences between the observed and the values expected by fractional differencing become more pronounced as short-range and seasonal parameters get introduced to the models, going all the way up to 1.22 (DFA) and 1.31 (smoothed spectral regression) in models with nested seasonal patterns, but also steadily increase as incidental short-range parameters are included in the simulations. While the GPH and Sperio estimate stay quite close to the expected values as incidental short-range parameters are included in the simulation models, Whittle's estimate does not, as it goes up to 0.61 in the ARIMA (1, 0, 0) case and to 1.01 for ARIMA (1, 0, 1). Whittle's estimate is known to absorb short-range

Table 4.2 Spectral Regression Estimates in Simulated Data With Varying Inputs for Short-Range or Long-Range Parameters: DFA, Smoothed Regression, GPH, Sperio, Whittle, and Fractional Differencing With and Without Covariates

ARFIMA (p, d, q) Simulation	Input Conditions	DFA H	Spectral Regression (Smoothed) $\hat{H}$	Expected Value d	GPH Estimator d	Sperio's Estimate d	Whittle's Estimate δ	Fractional Differencing d	Fractional Differencing With Covariates d
(0, 0, 0)		0.58	0.49	0.00	−0.09	−0.08	0.00	0.00	—
	$H = 0.65$	0.71	0.74	0.15	0.22	0.18	0.17	0.18	—
(0, d, 0)	$H = 0.75$	0.84	0.80	0.25	0.33	0.27	0.29	0.29	—
	$H = 0.85$	0.95	0.92	0.35	0.45	0.37	0.43	0.43	—
(1, 0, 0)	$\varphi = 0.70$	0.89	0.67	0.00	0.00	0.01	0.61	0.49	0.01
(1, 0, 1)	$\varphi = 0.70$ $\theta = -0.50$	0.94	0.73	0.00	0.01	−0.02	1.01	0.49	0.00
(1, 0, 1) × (1, 0, 0)	$\varphi = 0.70$ $\theta = -0.50$ $\Phi = 0.90$ Period =7	1.22	1.31	0.00	0.70	0.69	0.70	0.50	0.06
(1, 0, 1) × (1, 0, 0)	$\varphi = 0.70$ $\theta = -0.50$ $\Phi = 0.90$ Period = 12	1.01	1.26	0.00	0.93	0.80	1.07	0.50	0.05

Note. Expected values of d are based on the input conditions for the arfima simulations. Fractional differencing results are obtained using fracdiff for the estimates without covariates and those with nonseasonal short-range covariates. The arfima package was used to estimate d in the models with seasonal covariates. ARFIMA = autoregressive fractionally integrated moving average model; DFA = detrended fluctuation analysis; GPH = Geweke and Porter-Hudak.

dependencies into its estimates (Beran, 1994; Taqqu et al., 1995). GPH and Sperio turn out to be also highly sensitive to dependencies of the seasonal variety, as is Whittle's estimate.

Table 4.2 also shows that the results for fractional differencing stays close to the input values, provided that the modeled short-range and seasonal dependencies are included as covariates. In a way, this is a trivial result because of the correspondence between the method used to generate the simulated series and those used to generate their parameter estimates, but this result nonetheless shows the ability of fractional differencing methods to distinguish long-range fractal patterns from short-range patterns, seasonal and otherwise, that is absent from the other techniques.

The correlation between short-range and long-range dependency estimates is a recurring dilemma in the estimation of fractal patterns in stationary time series. Power spectra will distinguish the short range by showing a nonlinear pattern in the data points, whereas the long range displays a linear pattern, thus creating power spectra with distinct appearances (Stadnitksi, 2012a; Wagenmakers et al., 2004). Porter-Hudak (1990) achieves model improvements by building the seasonal cycles into the spectral regression equation, such that the estimation of d in Equation 4.20 corrects for the cycle that corresponds to the seasonal lags in the time domain. This approach deserves further development. However, at the time of this writing, the best response to this dilemma seems to be to use a fractional differencing approach with a competitive stepwise modeling (Wagenmakers et al., 2004), such that these sources of variability can be disentangled in a systematic manner, as was shown in Chapter 2.

C. The Hurst Exponent Revisited

The reader may have noted that one of the disturbing aspects of conducting the types of analyses described here is that the various techniques use different computations to generate the Hurst exponents. Table 3.2 in Chapter 3 shows small, and sometimes not-so-very-small, discrepancies in the results depending on whether this exponent is computed based on the differencing parameter, or based on the slope of the power spectrum. Another way of stating this result would be that there are discrepancies depending on whether the Hurst exponent is generated in the time domain or in the frequency domain. This issue extends to the techniques discussed in this chapter, as illustrated by the computations shown in Table 4.1, which show that when white noise and pink noise are simulated, the Hurst exponent is still quite close, whereas in case of the random walk, the Hurst exponents cannot be said to be similar at all. The estimation for the simulations shown in Table 4.2 are similarly discouraging. The discrepancy between the imputed values of H and those subsequently obtained show considerable variation, as well as an apparent dependency of H values on the inclusion of nonfractal short-range autocorrelation patterns in the simulations. The takeaway from these observations is, of course, that it is essential to report H in the analytical context in which it was obtained and adjust its interpretation accordingly. Moreover, additional simulation studies should help the field converge toward a unified approach to the estimation of the Hurst exponent and differencing parameters.

D. Chapter Summary and Reflection

One of the persistent dilemmas in fractal time series research is the fact the there are many different approaches, and up to this point, there is little consistency in the preferences of researchers where it comes to the analysis of fractal patterns. As mentioned in the previous section, this state of affairs complicates the estimation of the Hurst exponent, but it also introduced variability in how fractality is defined. The fractal variance approaches rely on variability found in the time series, and the estimation of its linear relationship to the size of the sample drawn from the original series, while spectral regression relies on the even distribution of nonrandom patterns across the frequency domain. These two phenomena are quite distinct and concluding fractality based on H or d says something very different about what goes on in these data, depending on the analytical approach that was taken. However, as in any fractal analysis, the remarks made in Chapter 1 about sample size requirements apply to the approaches discussed in this chapter as well, which is to say that the detection of patterns within patterns requires a large number of observations.

A major weakness about the approaches discussed in this chapter is their lack of ability to systematically account for the differences between short-range, seasonal, and fractal sources of variability. They are not distinguished in the parameters of the model, and a systematic assessment of their relative influence on the overall variability in the data is therefore difficult, as is illustrated most clearly in the simulation results shown in Table 4.2, in which those patterns are conflated in many instances. Clearly, there is a need for a large-scale simulation study to get a clearer sense of the conditions under which this conflation occurs in the estimation process using these techniques. In the meantime, the preliminary exploration of the time series and periodograms will be needed to decide whether the use of these approaches is indicated to distinguish random patterns from fractal patterns. It is also necessary to assume, in those cases, that no short-range dependencies are at play unless corrections are made to the input data, or the seasonal components can be built into the spectral regression equation (Porter-Hudak, 1990). Inspection of ACF plots, such as those shown in the figures in Chapter 2 will help determine what corrections are needed. And if short-range and seasonal dependencies seem to be at work, a stepwise fractional differencing approach is recommended to distinguish them from the fractal ones.

5

VARIATIONS ON THE FRACTALITY THEME

While the investigation of fractality may be used to address many areas of interest in complexity theory, there are other questions that also involve time series that it is decidedly ill equipped to address. One such question is the degree to which systemic turbulence can be predicted on the basis of small fluctuations at a previous point in time, also known as sensitivity to initial conditions, which is a primary concern of chaos theory. Chaos theory is a perspective that is also concerned with time series (Sprott, 2003), and therefore, it is of related interest to the discussion here. Below is a brief appreciation of the relevance of its insights. The reader may also have noticed that much of the work discussed in this book does not relate time series variables to any other variables. The question to what extent a process captured in a univariate time series is affected by the timing of the behavior of other systems or parts of systems is pertinent and often of importance to complexity theorists. Multivariate time series have been designed to address this issue and will also be briefly discussed below. In its focus on fractality, this book is concerned with long-range processes that are irregular and how they can be differentiated from regular short-range processes. It is also possible, of course, that long-range dependencies can be detected in a time series trajectory that are regular, for instance, if there is an annual cycle in a string of daily measurements. A brief discussion of the estimation of regular long-range patterns is also provided below. Last, time series trajectories can be affected by sudden changes in the circumstances affecting the systemic behavior of interest, such as specific interventions. The approaches to address such sudden impact are well covered in the existing methodological literature and will therefore be only briefly discussed in this chapter with an eye toward irregularity. These related techniques offer great opportunities to expand the analysis for fractal patterns to include situations of extreme turbulence, sets of time series measuring behaviors that are closely related, and behaviors that change in response to specific events. Furthermore, they offer a framework to distinguish regular from irregular long-range dependencies.

A. Sensitive Dependence on Initial Conditions

Time series can be irregular in many different ways, and the analytical approaches presented here provide only partial coverage of the scenarios we know. Among the most well-known of such irregular patterns are those proposed by chaos theory (Poincaré, 1914/2012). Chaos theory seeks to explain large-scale turbulence in a system in terms of infinitesimal fluctuations at an earlier point in time such that those fluctuations gradually evolve the system into a state of massive disruption. This process is referred to in the literature as *sensitivity to initial conditions* to refer to the small things that can produce these big results. The proverbial example of sensitivity to initial conditions is the Brazilian butterfly that flaps its wings to produce a hurricane later on. While the empirical status of this metaphor is questionable, it is nonetheless very telling in that it illustrates how very small fluctuations at one point in time can evolve into turbulence.

The work of MIT (Massachusetts Institute of Technology) meteorologist Edward N. Lorenz on weather prediction in the mid-1950s was of seminal importance to the development of chaos theory. Skeptical about the use of linear time series models, such as those discussed by Box and Jenkins (1970) and at the beginning of Chapter 2 of this book, to predict long-range weather patterns, Lorenz was searching for models with greater flexibility to describe nonlinear patterns. He noted that small rounding errors in the input values for simulation models produced dramatically different forecasting trajectories. This discovery resulted in the formulation of models that partition the variance components such that apparent random weather patterns can be described through a relatively simple set of differential equations (Lorenz, 1963). One of the lasting realizations emanating from Lorenz's work is that weather trajectories (e.g., temperature, precipitation) are not as predictable in the long term as is presumed under the linear statistical models that were used at the time (Palmer, 2008), and they require the modeling of irregular patterns over and above the seasonal ones, a concern that is also addressed in this book. However, the time series models used to describe chaos are deterministic functions, and while these are complex and interesting, they are not well suited to fit the types of data discussed in this volume, because measurement error would overwhelm the intricacies of the processes postulated by chaos theory. For further information about these deterministic functions, Kaplan and Glass (1995) and Sprott (2003) offer excellent introductions.

B. The Multivariate Case

The analyses discussed in this book have in common that they concern univariate data sets, where the outlook of the time series trajectory is modeled and predicted strictly on the basis of previous observations on the same variable, making this approach particularly suitable for the estimation of long-range fractal patterns. However, researchers are sometimes more interested in how a given time series is affected by other variables, because they want to know what the contingencies are for particular types of long-range behavior. These contingencies may be time series as well, such that the contingent series and the behavioral outcome series are coupled and the correlations between the two series can be analyzed in a manner that is sensitive to the time of occurrence of the two behavioral components of the relationship. An example of a situation where such an approach would be suitable is when behavioral synchronization is studied—that is, a nonrandom coupling of behavior of individuals over time. Tschacher (2019) analyzed such coupling patterns to determine the extent to which therapists and their clients are synchronizing gestures, facial expressions, and other instances of nonverbal communication, and he assesses whether more synchronization is associated with better therapeutic outcomes. In a related study, he also studied the time-dependent coupling between physiological measures of interacting spouses (skin conductance) to determine the extent to which behavioral synchrony differed depending on whether the couples were in cooperative or conflict situations. (There was greater synchrony in conflict situations; Tschacher, 2019.)

Likewise, Molenaar et al. (2009) collected time series of positive–negative affect measurements in the interactions between fathers/stepfathers and sons or stepsons. There were 80 recordings across the time scale. Eight biological sons, eight stepsons, and their (step)fathers participated in the study. It was determined to what extent there was covariation in the affect responses of participants in the course of the interactive episodes. The researchers found that anger and involvement in both interactive partners were related in a time-dependent manner. An increase in anger in one individual was correlated with an increase in involvement of the other part of the dyad in subsequent measurements, thus indicating how anger triggers an involvement response in the dyadic partner that reverberates across the time spectrum. While the connection of long-range time series to other long-range time series for the analysis of causality as a time-dependent phenomenon is an area of great promise, it falls outside of the scope of this book. The interested reader might consult the VAR package in R. Other approaches include the group iterative multiple model estimation (GIMME, Gates & Molenaar, 2012), which utilizes a structural equation framework to estimate the interrelationships between trajectories, allowing for a distinction between individual and group-level information. A more detailed discussion of this approach can be found in Molenaar (2015).

C. Regular Long-Range Processes and Nested Regularity

The concern of this book is the estimation of irregular long-range patterns. These are, of course, not the only type of long-range patterns one might encounter in time series data. Long-range patterns can also be regular. An example of such regular long-range cycles would be annual dependencies in trajectories of observations recorded daily. Our ability to handle such long-range dependencies statistically has been significantly enhanced with the development of the *Trigonometric Box–Cox ARMA Trend Seasonal (TBATS)* model. In brief essence, TBATS conducts a spectral analysis based on periodicities provided by the investigator. TBATS has two major assets. One is that it can model the long range with relative ease—for example, 365.25 lags for an annual cycle in daily recordings. Second, it can handle nested periodic structures—for example, weeks (7 lags) within years (365.25 lags). While the specification of the seasonal lags of the model is within control of the investigator, any remaining ARMA processes are mechanically corrected, and the TBATS output in R specifies at which lags autoregression and moving average patterns were found. For lucid introductions to the technique, refer to DeLivera and Hyndman (2012) and Hyndman and Athanasopoulos (2012). Koopmans (2020) applied TBATS to daily high school attendance data to show how weekly and annual cycles can be detected in those data.

The diagnostic process for detecting long range and nested regular cycles in time series data is not much different from that for short-range seasonal cycles. Figure 5.1 shows the ACF plots for the births to teens data at 50 and 500 lags. It can be seen that the regular cycles are quite pronounced both in the short range (weeks) and in the long range (years). These plots can be used as a justification for the use of TBATS to estimate the impact of

Figure 5.1   Short- and Long-Range Nonrandom Patterns in the Births to Teens Data: Autocorrelation Function (ACF) Plots at 50 and 500 Lags

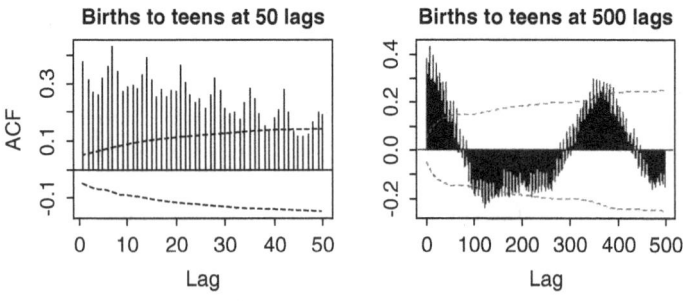

short- and long-range cycles within a single analytical model. For example, one could take a stepwise approach and compare the goodness of fit of a weekly model to a weeks-plus-years model to determine whether the annual dependencies contribute to the overall variability in the model over and above that accounted for by the weekly model (see Koopmans, 2020, for an example of this approach). Parenthetically, the high degree of weekly regularity in births to teens data shown in this figure has been attributed to the practice of physicians to induce labor early in the week so that as much as possible, the deliveries take place when staff and resources are at their highest levels (P. Hamilton, personal communication, May 8, 2018; Restrepo et al., 2018).

D. The Impact of Interventions

Pragmatic disciplines such as education, medicine, and nursing are often interested in measuring the impact of certain interventions on behavior, as exemplified by the use of randomized controlled trial studies to compare outcomes in groups that have been exposed to different treatment regimens (e.g., Shadish et al., 2002). Experimental designs can also be used in the context of time series, where frequently repeated observations of a single case provide a firm baseline against which to evaluate the impact of certain policy initiatives. Hence, these approaches are often referred to as inter-rupted time series or intervention analysis (McDowall et al., 1980). One common use of this design is the behavioral modification studies in educa-tional settings, where it can be assessed whether certain reinforcement programs can trigger desirable behaviors, or reduce undesirable ones, in classroom settings, playroom settings, or elsewhere. For instance, Hall et al. (1971) examined the impact of teacher reinforcement (praise and games) in response to students raising hands before talking on talking out behaviors in an elementary school classroom. Fifty classroom sessions were observed, with reinforcements being introduced at the 20th classroom session, with-drawn at the 40th session to be reintroduced in Session 45. The investiga-tors show a clear pattern in which the reinforcement program results in significant reductions in talking out behaviors, while this effect dissipates as reinforcement is withdrawn. Due to the frequent repetition of the meas-urements, this result allows for a causal attribution of the reduction in talk-ing out behaviors to teacher reinforcement of hand raising through praise and games.

This interventional analysis approach can be extended to the analysis of fractal patterns, presuming that a sufficient number of observations is included in the data set. Koopmans (2018a) analyzed the impact of the conversion of a large transfer high school into a smaller one on daily

82

attendance rates. The school originally had approximately 900 students, and that number was reduced to around 250. This school was and is a transfer high school, which is to say that it serves students who previously dropped out and now make their return to the educational system. Attendance patterns are not necessarily typical in this school, but its status as a transfer school makes it a particularly interesting case to evaluate the degree to which attendance rates are stable or not, and whether the reduction of the size of the school stabilizes their attendance rates. Figure 5.2, originally published in Koopmans (2018a), shows the effect of the size reduction on the stability levels in attendance rates. It can be seen in the figure that the preintervention trajectory is volatile and resembles Brownian motion, while the postintervention trajectory shows much less perturbation. In fact, the ADF test results for the two halves of the trajectory indeed do indicate that the series is nonstationary prior to the intervention (ADF = −3.15, p > .01), whereas the nonstationarity hypothesis is rejected for the postintervention trajectory (ADF = −4.57, p < .01). It can be concluded that the size reduction has had a stabilizing impact on the attendance rates in these schools. It is interesting to note that an upward pulse in attendance rates at the time of the intervention does not persist, but that gradual increases materialize in the longer range of the postintervention series (see Koopmans, 2018a, for more details).

It is also instructive in this context to take a closer look at the presidential approval ratings reported in Figure 1.1, where the effects of electing a new U.S. president or reelecting the incumbent president can be captured within an intervention analysis framework as well. This can be appreciated by looking at the dramatic increase in presidential approval ratings in 1974

Figure 5.2 The Effect of Reducing School Size: Pre- and Postintervention Attendance Rates

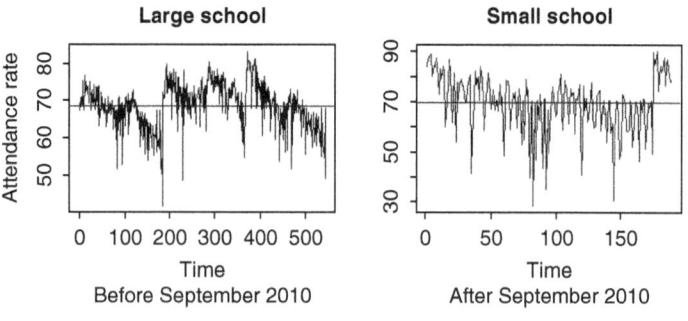

once Gerald Ford took over the presidency from Richard Nixon. Likewise, a drastic improvement in the approval ratings can be seen at the beginning of Ronald Reagan's term in 1981. These results illustrate how the shape of a time series trajectory can be associated with the timing of certain events, such as perhaps the Watergate scandal that unfolded in the early 1970s.

The effective use of intervention analysis can be expanded. In addition to the questions conventionally addressed through these approaches, such as the impact of timing on the change of certain outcomes, and the question whether those changes persist over the long range of the series, it can also be decided whether fractal patterns are induced through interventions in a system. It is often argued (e.g., Mandelbrot, 1997; Stadnitski, 2012a) that fractality could be an indicator of adaptability within the system, such that fluctuations may depend on changing circumstances, rather than on the self-perpetuation of rigid seasonal patterns. Whether such adaptive behavior is good or not good is a metaphysical question. The concern here is that intervention analysis can be used as a tool to determine whether fractality can be introduced or removed from a time series by acting on the system whose behavior is analyzed. In the aforementioned attendance study, the attendance rates go from volatile to fractal, which arguably is an improvement from a high degree of unpredictability. A higher degree of predictability in attendance is a beneficial outcome, then, in this context, if only for purposes of planning, educational administration, and classroom instruction.

6

CONCLUSION

As in any time series analysis, there are generally three reasons why one would like to estimate fractal patterns. The first is to make statistical corrections for unwanted sources of nonrandom variability, the second is to estimate in what way given observations in a time series depend on previous observations, and the third is to improve forecasting the behavior of interest into the future with greater reliability. The techniques described in this book are pertinent to each of these three cases, although the focus of the book is on the detection of fractality as a phenomenon of substantive interest. Ever since Mandelbrot's (1997) early work in this area, fractals have spoken to the imagination of many because they provide models for a built-in self-similarity in nature and culture that may, in turn, point to complex adaptive processes. The similarity may tell us something about the nature of the adaptive process and its manifestation, and raises the question how to capture the variability in those processes most effectively. There is a difference between geometric fractal shapes, which display a self-similarity that is strictly mathematically predictable, and the fractality that may occur in an empirically derived time series, where the self-similarity is a matter of appearance, rather than computational iteration. And since it is an empirical phenomenon, it is of scholarly interest.

The estimation of fractality in time series is clearly in its infancy, and it is based on work that was published in the early 1980s up to the mid-1990s (e.g., Beran, 1994; Granger & Joyeux, 1980). This work put most of the fundamentals in place according to which we still conduct these types of analysis. The data examples offered throughout this book show how useful it is to estimate long-range patterns. As far as we know, these cannot be captured any way other than by using the statistical techniques described in this book, and the analysis of time series of this length cannot be conducted reliably without estimating the fractal patterns. While there is great variability in how different researchers approach the question of fractality in the data, it seems that an approach that combines stepwise modeling within the fractional differencing framework with the generation of power spectra to confirm fractality seems to carry the greatest promise for the future.

Many issues remain unresolved at this point. For one, further work needs to be done to establish the reliability of the various estimation procedures and to enhance our capacity to disentangle the short-range and long-range dependencies in time series data, particularly in the frequency domain. In addition, greater uniformity needs to be attained in the estimation of the Hurst exponent, with perhaps one preferred approach each within the time domain and the frequency domain, instead of several. We are in a position, however, to evaluate the advantages and disadvantages of estimating fractality in the time domain versus the frequency domain, and particularly of fractional differencing versus PSDA. It may be helpful to summarize the main points of this evaluation.

A. Benefits and Drawbacks of Fractal Analysis

An undisputable strength of fractional differencing is its capacity to conduct comparative modeling such that the effects of including or excluding fractional estimates into the model can be estimated and that the relative impact of fractional and regular patterns on variance reduction can be systematically evaluated. This capability makes it possible to disentangle the correlated impact of long-range and short-range processes on the variability in the series. It also makes the approach particularly suitable for confirmatory research that relies on the estimation of the statistical significance of the difference of observed values from hypothesized ones. Thus, rejecting the null hypothesis that $d = 0$ provides evidence that the behavior of interest shows persistence or antipersistence, and a comparison of the goodness-of-fit indicators (BIC and LBQ) of competing models can be used to determine the significance of this effect *over and above* the short-range and regular dependencies that may be present in the trajectories (Wagenmakers et al., 2004). Conceptually, this approach is similar to a conventional stepwise regression analysis, with the exception that the predictors whose efficacy is evaluated concern previous measurements of the same variable, rather than concurrent measurements of a different one. PSDA, fractal variance estimation, and spectral regression are not similarly equipped to make systematic comparisons between models containing a varying combination of within-subject predictors. Moreover, with the exception of the estimation of the slope of the power function, these techniques are nonparametric (i.e., there are no assumptions about the distribution of the parameters in the population). Therefore, they are less suitable for confirmatory purposes.

On the other hand, the notion that you can infer fractality from the differencing parameters in a time series is an approach that is not without its detractors. While we can decide conclusively whether there is persistence in a time series data set, opinions might differ about what this means, other than that there is long-range nonrandomness in the series that needs to be statistically controlled for. The great advantage of power spectra is that they actually demonstrate fractality by showing the scale invariance of those nonrandom patterns over the long range. This invariance shows up in a strictly linear relationship between log power and log relative frequency, a relationship that shows that dependencies show up in the series irrespective of the scale at which those patterns are described (Bak et al., 1987; Mandelbrot, 1997). Therefore, power spectra can be used to fortify a fractal interpretation of the fractional differencing results. In many instances, it may be useful, then, to use fractional differencing and PSDA in conjunction, because it couples a strong confirmatory process with a more detailed description of the behavior of the data points to support the conclusion of fractality in the data. To explore fractal patterns in a time series, it may also be helpful to generate plots that are suggestive of self-similarity, such as those shown in Figure 1.3. Deciding on the scaling of these patterns in the visual plots by varying the length of the truncated series such that the patters would repeat is a trial-and-error process, and thus, it requires close scrutiny of the raw data points. Such scrutiny is, of course, a prerequisite for most effective statistical modeling, irrespective of whether we are looking for fractal patterns or for something else.

A drawback of fractional differencing is that it requires us to assume that time series are stationary, which is not always the case. In cases where they are not, it typically relies on the analysis of stationary increments. While this approach is effective for the estimation of short-range patterns, volatility and metastability will mostly disappear from the trajectory. Thus, nonfractional differencing results in a loss of authenticity of the time series. This point can be appreciated by comparing nonstationary data to their stationary increments (see Figure 2.3 and the second and third rows of Figure 2.5).

A strength of PSDA is that it provides a single analytical framework that enables us to handle both stationary and nonstationary cases. Power spectra can be generated regardless of the stationarity of the time series, and the slope of these spectra can be used to draw conclusions about the extent to which these series are volatile. Thus, the approach brings out the continuity of the relationship between stability, metastability, and volatility, as the slopes of the power spectra of these scenarios gradually change. This continuity between the stationary and nonstationary cases has been best articulated by Eke et al. (2000), whose article links the appearance and clustering of the time series to the appearance of the power spectra and the interpretation of the Hurst exponents and the slopes of those power spectra. Here, this continuity can be seen in the power spectra and their slope values reported for simulated white noise, pink noise, and the random walk in Figure 3.3, and in the time series for these simulations shown in Figures 2.1, 2.2, and 2.5, which show a gradual shift from randomness to volatility by way of short-range autoregression and metastability.

However, it has also been noted above that the slope of power spectra is not readily interpretable unless the relationship between log power and log relative frequencies is linear. In other words, the slopes of power spectra estimate the degree to which time series display nonrandomness and scale invariance. Short-range autoregression does not fit onto that continuum, because there typically is no linear relationship between these two variables, rendering these slopes uninterpretable. The continuity between short-range autoregression is therefore better captured in the time domain, as is illustrated by our ability to meaningfully compare models estimating across these two premises.

Complexity researchers typically prefer to use the Hurst exponent as a way to determine whether time series are stable or not. However, this exponent does not capture a continuity between randomness and volatility in a straightforward manner. Within a fractional differencing framework, H expresses a range from antipersistence to persistence in a stationary series on a scale from 0 to 1, and equals 0.5 in the absence thereof. In PSDA, the exponent indicates fractality in the series irrespective of its stationary status, and as per Eke et al.'s (2000) discussion, the computation of H depends on whether the slope values fall within the range of fGn or within the range of fBm. Thus, its range indicates fractality separately for these two scenarios and defines a range from 0 to 1 within each of these two scenarios. It is helpful to realize that the determination of fGn or fBm in the frequency domain often, but not necessarily, correspond to the findings of the stationarity test in the time domain, such as those shown in Table 2.1.

The real data results discussed in this book illustrate how fractional differencing and PSDA enable certain estimations but not others. While the Hurst exponents and the periodograms indicate long-range dependencies in all data sets, the power spectra do not support this conclusion for the births to teens data. Likewise, the competitive modeling strategy reveals that there is a fractional effect over and above the short-range estimates in the political orientation and school attendance data, whereas in the births to teens data, there is not. Fixing the differencing parameter to 1 in this latter case disables the estimation of Hurst within the time domain unless DFA or Higuchi is used. The nonlinear pattern in the power spectrum for the births to teens data similarly undermines the estimation of the Hurst exponent in the frequency domain. In conjunction, these observations support the case for an integrated approach that capitalizes on the strengths of each to obtain the most thorough assessment of the long-range dependency in time series data. However, if researchers insist on using the Hurst exponent to answer the question of fractality, it is important to be reminded that results may differ depending on the estimation method that is being used.

Many dynamical researchers complain about the lack of consistency of the results when estimating the Hurst exponent, and the dependency of those results on the analytical angle of incidence and programming routines used. For example, given the variability in H and d estimates shown in the simulated spectral regression results in Table 4.2, one may wonder to what extent these varying estimates are really measuring the same thing. This is clearly not the case in those instances where some measures, such as Whittle's estimate absorb short-range dependencies while others do not, and yet others (GPH and Sperio) seem to absorb seasonal dependencies but show no sensitivity to incidental short-range autocorrelations. This situation calls for measures that are capable of distinguishing those scenarios, particularly if the dependencies under investigation have substantive import, rather than merely being modeled for purposes of statistical correction. As argued repeatedly in this book, the case for such substantive relevance can often be made for the distinction of seasonal and fractal patterns, since the pragmatic or policy implications of them may be very different. There is a need for greater coherence in this area, and for a more outspoken preference among scholars about which analytical approaches yield the most reliable estimates, and those most readily interpretable in terms of the research questions. The empirical work discussed in this book appears to benefit most from the aforementioned approach that integrates the comparative modeling capabilities of fractional differencing with the substantive meaning imparted by the power spectra with linear trends in the data points. The generalization of this impression to other data awaits rigorous future simulation studies.

B. Interpretation of Parameters in Terms of Complexity Theory

We indicated previously that there is evidence of self-similarity or fractality in the daily attendance ratings in School 1, unemployment in the United States, and in the left–right political orientation of CDA and PvdA voters in the Netherlands. What exactly does that mean? There would be no fractality in these data if the estimates of long-range dependencies were nonsignificant, but if they are significant, does that necessarily imply fractality? Based merely on finding of persistence or antipersistence in the series, the best we can say is that fractality cannot be ruled out. As noted above, our ability to fit power law functions strengthens the fractal interpretation of these data because they show in each of these instances how, irrespective of the range of relative frequencies considered, the same linear pattern is observed, thus showing the self-similar character of these data. The findings for attendance and political orientation stand in contrast with the births to teens data, where regular patterns prevail, and the fitted power law functions do not reveal self-similarity.

Self-similarity is a phenomenon of interest to dynamical scholars because it points to the adaptive behavior of the system of interest (Stadnitski, 2012a), which is to say that it suggests that the system maintains its behavioral identity (i.e., the patterns repeating), in light of changing circumstances (the varying size of the time frames within which these patterns repeat themselves). Obviously, a univariate description only captures a small aspect of the behavior of a system, which is likely to be predicated on a wide range of causal factors, whose interplay produces these fractal patterns. In the case of school attendance, for instance, the fluctuations in attendance are likely to be related to varying levels of parental support, instructional effectiveness, school building leadership, and perhaps economic factors, public transportation, and weather patterns—factors that, in conjunction, may be responsible at least in part for the kind of fluctuations seen in these time series.

Finding Hurst exponents of $H \neq 0.5$ has also been linked sometimes to other systems properties, such as *self-organized criticality*, a state in the system where repeated exposure to the same input conditions ultimately results in a qualitative transformation of the system (Bak, 1996). A prototypical example of this phenomenon is Bak's sandpile, which is poured out on a flat surface, such that the ongoing addition of grains produces an avalanche at certain points due to the mounting friction between the grains in the pile. Perhaps the metastability that is observed in the left–right orientation in CDA and PvdA voters in the Dutch electorate is a precursor to the unexpected proliferation of anti-immigrant parties later on, as the electorate suddenly changes party affiliation on a large scale in the face of repeated disappointing experiences with the political system. Similarly, lengthy stationary processes may have preceded the political shifts observed at the time of this writing in the United States (the election of Donald Trump), the United Kingdom (the Brexit vote), and some other countries (e.g., Poland, Hungary).

C. A Note About the Software and Its Use

The work reported in this book relies heavily on the use of R, the open source statistical software package. The value of R to the field is hard to overestimate. It is available to anyone who has an internet connection and wishes to conduct statistical analyses, and it is a relatively friendly programming environment. New packages are developed all the time, containing important (and not so important) statistical innovations, including those discussed in this book. R is also a great asset to independent researchers who cannot rely on the purchasing power of institutions to obtain the standard statistical packages such as SAS, IBM-SPSS, Stata, and others. Thus, R plays an instrumental role in the democratization of the realm of statistics. There is by now also a very extensive body of documentation on the use of R, much of it accessible from within the programming environment itself, and many of the major academic publishers, Sage included, have put out statistical texts that explicitly discuss R applications.

R is not the only package that enables the analysis of fractal patterns. The SAS PROC IML routine has a set of ARFIMA subroutines and Stata Version 14 offers fractional differencing options within the time series menu. MAT-LAB and Mathematica also have the capacity to conduct fractional differencing analyses and generate power spectra. While IBM-SPSS Version 25 has considerable time series analysis capabilities, those analyses require the separate purchase of the Forecasting add-on feature, which, unfortunately, does not do fractional differencing. Conducting spectral analysis, on the other hand, is part of the standard features of most packages, including SPSS. To the benefit of the field of fractal research, Stadnitski (2012b) shows how to generate power spectra in SAS, SPSS, and R.

In the R environment, several packages are available to cover the analyses discussed in this book. The most important ones are `fracdiff`, `fractal`, `forecast`, and `arfima.sim`. In addition, the analyses described here rely on two general purpose time series packages, `TSA` and `tseries`. When estimating fractal patterns, the coefficients and goodness-of-fit statistics generated by `fracdiff` and those produced in Stata are largely consistent. In the selection of short-range parameter estimates, `fracdiff` is remarkably inflexible, as they can only be added in a cumulative manner. In Stata, on the other hand, one can estimate, say, autocorrelations at two lags while leaving the first lag out of the estimation. Particularly, when seasonal patterns (5, 7, or 12 lags) are considered, the need to estimate all preceding lags as well is impractical to the point of being prohibitive.

Neither Stata nor the packages in R we discuss in the book permit the estimation of the differencing parameter when multiplicative seasonal models are fitted. In Stata, one can estimate seasonal lags by themselves without having to also estimate the intermediate lags. While this offers a workaround to the estimation of seasonal and fractal patterns in conjunction, it may result in an underestimation of seasonal patterns, for which seasonal nesting models produce better results (Cryer & Chan, 2008). In R, the option of estimating fractality in multiplicative models is available in the `arfima` package, thus making it possible to evaluate the relative influence of seasonal and fractal patterns very systematically. However, the computation of the goodness-of-fit statistics in that package is idiosyncratic and is not readily translatable into those used by other time series packages. For this reason, arfima is not used for the present data analyses, unless the joint estimation of seasonal and fractal patterns is indispensable, such as in some of the entries in Table 4.2. The `arfima.sim` routine, on the other hand, produces excellent results and has been used for all simulations.

Both conceptually and computationally, accounting for seasonal and fractal patterns in conjunction is an underdeveloped field. This state of affairs is probably attributable, at least in part, to the fact that much of our capacity to analyze fractal patterns was developed in areas where seasonality is not of any obvious interest, such as the analyses of irregular hearbeat (Eke et al., 2000), depth perception (Aks & Sprott, 2003), or political left–right orientation (Eisinga et al., 1999). Yet many of the data discussed in this book, such as the births to teens and school attendance, attest to the importance of estimating fractal effects over and above the seasonal ones, and therefore, we need a better integration of our statistical programming capabilities in those areas.

Many of the most common computational routines for the analysis of fractal time series analysis are covered in this book. The appendix to this book (available on the accompanying website for this book at **https://study .sagepub.com/researchmethods/qass/koopmans-using-time-series**) offers access to the data sets, both the simulated and the real ones, and it contains the programming statements that were used to conduct the analyses reported and to produce the figures. There is also a section in the script for Chapter 2 describing how fractional differencing is done in Stata. The reader is encouraged to try these routines to conduct the fractal analyses and replicate the results described in these chapters, as well as replicate the figures.

Note also that *R-Studio* is an interface that greatly facilitates the use of R, and like R, it is free, so its use is highly recommended as well.

REFERENCES

Adriani, P., & McKelvey, B. (2007). Beyond Gaussian averages: Redirecting international business management research toward extreme events and power laws. *Journal of International Business Studies, 38*(7), 1212–1230. https://doi.org/10.1023/A:1021431631831

Aks, D. J., & Sprott, J. C. (2003). The role of depth and 1/f dynamics in perceiving reversible figures. *Nonlinear Dynamics, Psychology, and Life Sciences, 7*, 161–180. https://doi.org/10.1023/A:1021431631831

Bak, P. (1996). *How nature works: The science of self-organized criticality.* Springer. https://doi.org/10.1007/978-1-4757-5426-1

Bak, P., Tang, C., & Wiesenfeld, K. (1987). Self-organized criticality: An explanation of *1/f* noise. *Physical Review Letters, 59*(4), 381–384. https://doi.org/10.1103/PhysRevLett.59.381

Barabási, A. L. (2014). *Linked: How everything is connected to everything else and what it means for business, science and everyday life.* Basic Books.

Beran, J. (1994). *Statistics for long-memory processes.* Chapman & Hall.

Bloomfield, P. (1976). *Fourier analysis of time series: An introduction.* John Wiley.

Borgatti, S. P., Everett, M. G., & Johnson, J. C. (2013). *Analyzing social networks.* Sage.

Box, G., & Jenkins, G. (1970). *Time series analysis: Forecasting and control.* Holden-Day.

Box-Steffensmeier, J. M., & Smith, R. M. (1998). Investigating political dynamics using fractional integration methods. *American Journal of Political Science, 42*(2), 661–689. https://doi.org/10.2307/2991774

Brandt, P. T., & Williams, J. T. (2006). *Multiple time series models* (Sage University Papers on Quantitative Applications in the Social Sciences, Vol. 148). Sage.

94

Brown, C., & Liebovitch, L. (2010). *Fractal analysis* (Sage University Papers on Quantitative Applications in the Social Sciences, Vol. 165). Sage. https://doi.org/10.4135/9781412993876

Caccia, D. C., Percival, D., Cannon, M. J., Raymond, G., & Bassingthwaighte, J. B. (1997). Analyzing exact fractal time series: Evaluating dispersional analysis and rescaled range methods. *Physica A, 246*(3–4), 609–632. https://doi.org/10.1016/S0378-4371(97)00363-4

Chen, Y., Ding, M., & Kelso, J. A. (1997). Long memory processes ($1/f^\alpha$ type) in human coordination. *Physical Review Letters, 79*(22), 4501–4504. https://doi.org/10.1103/PhysRevLett.79.4501

Clarke, H. D., & Lebo, M. (2003). Fractional (co)integration and governing party support in Britain. *British Journal of Political Science, 33*(2), 283–301. https://doi.org/10.1017/S0007123403000127

Cryer, J. D., & Chan, K. S. (2008). *Time series analysis: With applications in R* (2nd ed.). Springer. https://doi.org/10.1007/978-0-387-75959-3

Delignières, D., Fortes, M., & Ninot, G. (2004). The fractal dynamics of self-esteem and physical self. *Nonlinear Dynamics, Psychology, and Life Sciences, 8*, 479–510.

Delignières, D., Torre, K., & Lemoine, L. (2005). Methodological issues in the application of monofractal analysis in psychological and behavioral research. *Nonlinear Dynamics, Psychology, and Life Sciences, 9*, 435–461.

De Livera, A. M., Hyndman, R. J., & Snyder, R. D. (2012). Forecasting time series with complex seasonal patterns using exponential smoothing. *Journal of the American Statistical Association, 106*(496), 1513–1527. https://doi.org/10.1198/jasa.2011.tm09771

Eisinga, R., Franses, P. H., & Ooms, M. (1999). Forecasting long memory left-right political orientations. *International Journal of Forecasting, 15*(2), 185–199. https://doi.org/10.1016/S0169-2070(98)00064-8

Eke, A., Hermán, P., Bassingthwaighte, J. B., Raymond, G. M., Perival, D. B., Cannon, M., Balla, I., & Ikrényi (2000). Physiological time series: Distinguishing fractal noises from motions. *Pflügers Archive: European Journal of Physiology, 439*, 402–415. https://doi.org/10.1007/s004249900135

Eliason, S. R. (1993). *Maximum likelihood estimation: Logic and practice. Quantitative applications in the social sciences* (Vol. 96). Sage. https://doi.org/10.4135/9781412984928

Feder, J. (1988). *Fractals.* Plenum Press. https://doi.org/10.1007/978-1-4899-2124-6

Gates, K. M., & Molenaar, P. C. M. (2012). Group search algorithm recovers effective connectivity maps for individuals in homogeneous and heterogeneous samples. *NeuroImage, 63*(1), 310–319. https://doi.org/10.1016/j.neuroimage.2012.06.026

Geweke, J., & Porter-Hudak, S. (1983). The estimation and application of long memory in time series models. *Journal of Time Series Analysis, 4*(4), 221–238. https://doi.org/10.1111/j.1467-9892.1983.tb00371.x

Granger, C. W. J., & Joyeux, R. (1980). An introduction to long-memory time series models and fractional differencing. *Journal of Time Series Analysis, 1*(1), 15–29. https://doi.org/10.1111/j.1467-9892.1980.tb00297.x (Reprinted in *Time series with long memory,* pp. 49–64, by P. M. Robinson, Ed., 2003, Oxford University Press)

Guastello, S. J. (1995). *Chaos, catastrophe and human affairs: Applications of nonlinear dynamics to work, organizations and social evolution.* Erlbaum.

Hall, R. V., Fox, R., Willard, D., Goldsmith, L., Emerson, M., Owen, M., Davis, R., & Porcia, E. (1971). Teacher as observer and experimenter in the modification of disputing and talking out behaviors. *Journal of Applied Behavior Analysis, 4*(2), 141–149. https://doi.org/10.1901/jaba.1971.4-141

Hamilton, P., Pollock, J. E., Mitchell, D. A., Vincenzi, A. E., & West, B. J. (1997). The application of nonlinear dynamics to nursing research. *Nonlinear Dynamics, Psychology, and Life Sciences, 1,* 237–261. https://doi.org/10.1023/A:1021831811907

Harris, J. K. (2014). *An introduction to exponential random graph modeling* (Sage University Papers on Quantitative Applications in the Social Sciences, Vol. 173). Sage. https://doi.org/10.4135/9781452270135

Higuchi, T. (1988). Approach to an irregular time series on the basis of the fractal theory. *Physica D, 31*(2), 277–283. https://doi.org/10.1016/0167-2789(88)90081-4

Hurst, H. E. (1956). The problem of long-term storage in reservoirs. *Hydrological Sciences Journal, 1*(3), 13–27. https://doi.org/10.1080/02626665609493644

Hurst, H. E. (1965). *Long-term storage: An experimental study.* Constable.

Hyndman, R. J., & Athanasopoulos, G. (2012). *Forecasting: Principles and practice* (2nd ed.). OTexts. https://otexts.org/fpp2/

Kaplan, D., & Glass, L. (1995). *Understanding nonlinear dynamics.* Springer. https://doi.org/10.1007/978-1-4612-0823-5

Kauffman, S. A. (1993). *The origins of order: Self-organization and selection in evolution.* Oxford University Press. https://doi.org/10.1007/978-94-015-8054-0_8

Kauffman, S. A. (1995). *At home in the universe: The search for the laws of complexity and self-organization.* Oxford University Press.

Keele, L., Linn, S., & McLaughlin Webb, C. (2016). Treating time with all due seriousness. *Political Analysis, 24*(1), 31–41. https://doi.org/10.1093/pan/mpv031

Knoke, D., & Yang, S. (2019). *Social network analysis* (3rd ed., Sage University Papers on Quantitative Applications in the Social Sciences, Vol. 154). Sage.

Koopmans, M. (2015). A dynamical view of high school attendance: An assessment of short-term and long-term dependencies in five urban schools. *Nonlinear Dynamics, Psychology, and Life Sciences, 19*(1), 65–80.

Koopmans, M. (2018a). Exploring the effects of creating small high schools on daily attendance: A statistical case study, *Complicity: An International Journal for Complexity and Education, 18*(1), 19–30. https://doi.org/10.29173/cmplct29352

Koopmans, M. (2018b). On the pervasiveness of long range memory processes in daily high school attendance rates. *Nonlinear Dynamics, Psychology, and Life Sciences, 22*(2), 243–262.

Koopmans, M. (2020). Using time series analysis to estimate complex regular cycles in daily high school attendance, *Fluctuation and Noise Letters, 19*(1). https://doi.org/10.1142/S0219477520500030

Koopmans, M., & Stamovlasis, D. (Eds.). (2016). *Complex dynamical systems in education: Concepts, methods and applications.* Springer. https://doi.org/10.1007/978-3-319-27577-2

Kwiatkowski, D., Phillips, P. C. B., Schmidt, P., & Shin, Y. (1992). Testing the null hypothesis of stationarity against the alternative of a unit root: How sure are we that economic time series have a unit root? *Journal of Econometrics, 54*(1–3), 159–178. https://doi.org/10.1016/0304-4076(92)90104-Y

Lauwerier, H. (1991). *Fractals: Endlessly repeated geometrical figures.* Princeton University Press.

Lorenz, E. N. (1963). Deterministic nonperiodic flow. *Journal of the Atmospheric Sciences, 20*(2), 130–141. https://doi.org/10.1175/1520-0469 (1963)020<0130:DNF>2.0.CO;2

Mandelbrot, B. B. (1997). *Fractals and scaling in finance: Discontinuity, concentration, risk.* Springer. https://doi.org/10.1007/978-1-4757-2763-0

Mandelbrot, B., & van Ness, M. (1968). Fractional Brownian noises and applications. *SIAM Review, 10*(4), 422–437. https://doi.org/10.1137/1010093

Maturana, H. R., & Varela, F. J. (1980). *Autopoiesis and cognition: The realizationoftheliving.* Reidel. https://doi.org/10.1007/978-94-009-8947-4

McDowall, D., McCleary, R., Meidinger, D., & Hay, R. A., Jr. (1980). *Interrupted time series analysis* (Sage University Papers on Quantitative Applications in the Social Sciences, Vol. 21). Sage. https://doi.org/10.4135/9781412984607

McKuen, M. B., Erikson, R. S., & Stimson, J. A. (1989). Macropartisanship. *American Political Science Review, 83*(4), 1125–1142. https://doi.org/10.2307/1961661

Molenaar, P. C. M. (2004). A manifesto on psychology as in idiographic science: Bringing the person back into psychology, this time forever. *Measurement, 2*(4), 201–218. https://doi.org/10.1207/s15366359mea0204_1

Molenaar, P. C. M. (2015). On the relation between person-oriented and subject-specific approaches. *Journal for Person-Oriented Research, 1*(1–2), 34–41. https://doi.org/10.17505/jpor.2015.04

Molenaar, P. C. M., Sinclair, K. O., Rovine, M. J., Ram, N., & Corneal, S. E. (2009). Analyzing developmental processes on an individual level using non-stationary time series modeling. *Developmental Psychology, 45*(1), 260–271. https://doi.org/10.1037/a0014170

Ostrom, C. W., Jr. (1990). *Time series analysis: Regression techniques* (2nd ed., Sage University Papers on Quantitative Applications in the Social Sciences, Vol. 9). Sage. https://doi.org/10.4135/9781412986366

Palmer, T. (2008). Obituaries: Edward Norton Lorenz. *Physics Today, 69*(9), 81–82. https://doi.org/10.1063/1.2982132

Peng, C. K., Buldyrev, S. V., Havlin, S., Simons, M., Stanley, H. E., & Goldberger, A. L. (1994). Mosaic organization of DNA nucleotides. *Physical Review E, 49*(2), 1685–1689. https://doi.org/10.1103/PhysRevE.49.1685

98

Peng, C. K., Mietus, J., Hausdorff, J. M., Havlin, S., Stanley, H. E., & Goldberger, A. L. (1993). Long-range anti-correlations and non-Gaussian behavior of the heartbeat. *Physical Review Letters, 70*, 1343–1346. https://doi.org/10.1103/PhysRevLett.70.1343

Pickup, M. (2014). *Introduction to time series analysis* (Sage University Papers on Quantitative Applications in the Social Sciences, Vol. 173). Sage.

Poincaré, J. H. (2012). *Science and method*. Dover. (Original work published 1914)

Porter-Hudak, S. (1990). An application of the seasonal fractionally differenced model to the monetary aggregates. *Journal of the American Statistical Association, 85*(410), 338–344. https://doi.org/10.1080/016214 59.1990.10476206

Prigogine, I., & Stengers, I. (1984). *Order out of chaos: Man's new dialogue with nature*. Bantam Books.

Reisen, V. A. (1994). Estimation of the fractional difference parameter in the ARIMA (*p, d, q*) model using the smoothed periodogram. *Journal of Time Series Analysis, 15*(3), 335–350. https://doi.org/10.1080/0162145 9.1990.10476206

Restrepo, E., Hamilton, P., Liu, F., & Mancuso, P. (2018). Relationships among neonatal mortality, hospital volume, weekly demand and weekend birth. *Canadian Journal of Nursing Research, 50*(2), 64–71. https://doi .org/10.1177/0844562117751313

Said, S. E., & Dickey, D. A. (1984). Testing for unit roots in autoregressive–moving average models of unknown order. *Biometrika, 71*(3), 599–608. https://doi.org/10.1093/biomet/71.3.599

Sayrs, L. W. (1989). *Pooled time series analysis* (Sage University Papers on Quantitative Applications in the Social Sciences, Vol. 70). Sage. https://doi .org/10.4135/9781412985420

Shadish, W. R., Cook, T. D., & Campbell, D. T. (2002). *Experimental and quasi-experimental designs for generalized causal inference*. Houghton Mifflin.

Shumway, R. H., & Stoffer, D. S. (2011). *Time series and its applications* (3rd ed.). Springer. https://doi.org/10.1007/978-1-4419-7865-3

Sowell, F. (1992). Modeling long-run behavior with the fractional ARFIMA model. *Journal of Monetary Economics, 29*(2), 277–302. https://doi .org/10.1016/0304-3932(92)90016-U

Sprott, J. C. (2003). *Chaos and time-series analysis*. Oxford University Press.

Stadnitski, T. (2012a). Measuring fractality. *Frontiers in Physiology, 3*, Article 127. https://doi.org/10.3389/fphys.2012.00127

Stadnitski, T. (2012b). Some critical aspects of fractality research. *Nonlinear Dynamics, Psychology, and Life Sciences, 16*, 137–158.

Stadnytska, T., Braun, S., & Werner, J. (2010). Analyzing fractal dynamics employing R. *Nonlinear Dynamics, Psychology, and Life Sciences, 14*(2), 117–144.

Stewart, I. (2012). *In pursuit of the unknown: 17 equations that changed the world*. Basic Books.

Sulis, W., & Combs, A. (1996). *Nonlinear dynamics in human behavior*. World Scientific. https://doi.org/10.1142/3173

Taqqu, M. S., & Teverovksy, V. (1997). On estimating the intensity of long-range dependencies infinite and infinite variance time series. In R. Adler, R. B. Feldman, & M. S. Taqqu (Eds.), *A practical guide to heavy tails* (pp. 177–217). Birkhauser.

Taqqu, M. S., Teverovsky, V., & Willinger, W. (1995). Estimators of long-range dependence: An empirical study. *Fractals, 3*(4), 785–803. https://doi.org/10.1142/S0218348X95000692

Tschacher, W. (2019, March). *Synchrony and embodiment: Empirical hypotheses derived from synergetics* [Keynote address]. International Nonlinear Science Conference, Coimbra, Portugal.

U.S. Department of Labor (2017). *Labor force statistics from the current population survey*. https://data.bls.gov/timeseries/LNS13000000

Van Orden, G. C., Holden, J. G., & Turvey, M. T. (2005). Human cognition and 1/f scaling. *Journal of Experimental Psychology, 134*(1), 117–123. https://doi.org/10.1037/0096-3445.134.1.117

von Berthalanffy, L. (1968). *General system theory: Foundation, development, application*. George Braziller.

Wagenmakers, E. J., Farrell, S., & Ratcliff, R. (2004). Estimation and interpretation of 1/f$^\alpha$ noise in human cognition. *Psychonomic Bulletin & Review, 11*(4), 579–615. https://doi.org/10.3758/BF03196615

Waldrop, M. M. (1992). *Complexity: The emerging science at the edge of order and chaos*. Simon & Schuster. https://doi.org/10.1063/1.2809917

Wang, Y., & Liu, Q. (2006). Comparison of Akaike's information criterion (AIC) and Bayesian information criterion (BIC) in selection of stock-recruitment relationships. *Fisheries Research, 77*(2), 220–225. https://doi.org/10.1016/j.fishres.2005.08.011

Wasserman, S., & Faust, K. (1994). *Social network analysis: Methods and applications*. Cambridge University Press.

Wijnants, M. L., Cox, R. F. A., Hasselman, F., Bosman, A. M. T., & Van Orden, G. (2012). Does sample rate introduce an artifact in spectral analysis of continuous processes? *Frontiers in Physiology, 3*, 495. https://doi.org/10.3389/fphys.2012.00495

Wong, A. E., Vallacher, R. R., & Novak, A. (2014). Fractal dynamics in self-evaluation reveal self-concept clarity. *Nonlinear Dynamics, Psychology, and Life Sciences, 18*(4), 349–369.

Yazawa, T., & Omata, S. (2019). *mDFA detects abnormality: From heartbeat to maternal vibration*. IntechOpen. https://doi.org/10.5772/intechopen.85798

APPENDIX

These files are available at **https://study.sagepub.com/researchmethods/qass/koopmans-using-time-series**.

Filename	File Type	Contents
AJPS2a	DAT File	Political attitudes, 1953–1993 ($N = 160$)
AJPS2	RDATA	Political attitudes, 1953–1993 ($N = 160$, formatted)
fracsim	RDATA	Simulations (white noise, short-range autoregression, pink noise, random walk)
fracsim1	RDATA	Seasonal simulations
fracsim2	RDATA	Simulations for Table 4.2
School Attendance	Stata Dataset	Attendance in School 1, 2010–2014 ($N = 735$)
teens	Stata Dataset	Births to teens, 1964–1966 ($N = 800$)
Unempl	Stata Dataset	U.S. unemployment, 1948–2017 ($N = 837$)
dweekly	Text Document	Left–right political orientation in the Netherlands 1978–1996 ($N = 988$)
School 1 Attendance	Text Document	Attendance in School 1, 2010–2014 ($N = 735$)
School 2 Attendance	Text Document	Attendance in School 2, 2007—2014 ($N = 1,245$)
Script Chapter 1	Text Document	Program lines for Chapter 1
Script Chapter 2	Text Document	Program lines for Chapter 2
Script Chapter 3	Text Document	Program lines for Chapter 3

Script Chapter 4	Text Document	Program lines for Chapter 4
Script Chapter 5	Text Document	Program lines for Chapter 5
teens1	Text Document	Births to teens, 1964–1968 ($N = 1,461$)
teens2	Text Document	Births to teens, 1964–1966 ($N = 800$)
US_Unempl	Text Document	U.S. unemployment, 1948–2017 ($N = 837$)

INDEX

ACF. *See* Autocorrelation function (ACF)

ADF test. *See* Augmented Dickey–Fuller (ADF) test

Akaike's information criterion (AIC), 30, 31

Amplitude of cycles, 44–46

Antipersistence, 22

Hurst exponent and, 40

ARFIMA. *See* Autoregressive fractionally integrated moving average (ARFIMA)

ARIMA. *See* Autoregressive integrated moving average (ARIMA)

ARMA (*p, q*) model, 16, 19

Attendance rates, 24

comparative modeling, 33–35, 34 (table)

fractional differencing, 29, 29 (table)

interventional analysis approach to, 81–82

periodogram and power spectra, 53, 54 (figure), 55

self-similarity in, 10–11, 10 (figure)

spectral density analysis, 53–55

stationarity tests, 26–29

time series and ACF plots, 25 (figure), 26

Augmented Dickey–Fuller (ADF) test, 21–22

for real data, 26–29

Autocorrelation, 2–3

differencing parameter and, 22

LBQ test for, 30–31

removal of, 30–31

Autocorrelation function (ACF), 15–16

pink noise, 23, 24 (figure)

random variability, 17, 17 (figure)

random walk, 19–20, 20 (figure)

of real data, 25–26, 25 (figure), 31

residual time series, 38, 39 (figure)

seasonal series, 19, 19 (figure)

short-range autoregression, 17 (figure), 18

Autoregression, 16

seasonal, 18

See also Short-range autoregression

Autoregressive fractionally integrated moving average (ARFIMA), ix, 12–13, 15, 23, 71–73

See also Fractional differencing

Autoregressive integrated moving average (ARIMA), 21, 44

goodness-of-fit measures for, 30

residuals of, 38

Backshift operator, 22

Bayesian information criterion (BIC), 30, 31

Births to teens in Texas. *See* Teen pregnancies in Texas

Brown, Robert, 20

Brownian motion, 20–21, 53

fractional, 55–57

periodogram and power spectra, 50–52, 51 (figure)

Chaos theory, 77, 78

Characteristic equation, 22

Competitive modeling strategy, 30–36, 88

Complexity, 9–12
Complexity theory, 9, 89–90
Curve length, 66–67

Daily attendance rates. *See* Attendance
    rates
Detrended fluctuation analysis (DFA),
    60–63, 67, 69 (table)
Discrete Fourier transform, 46–47
Dynamical theories, 6

Endogenous processes, 11–12
Euler constant, 70
Exogenous processes, 11–12

fBm. *See* Fractional Brownian
    motion (fBm)
fGn. *See* Fractional Gaussian noise (fGn)
Fourier transform, 46–49
Fractal analysis
    benefits and drawbacks, 86–89
    fractional differencing, 15–42
    interventional analysis approach,
        81–83
    PSDA, 43–58
Fractality (fractals), 7–8, 77–83
    definition, 1, 41
    dimension, 60, 65–69
    by fractional differencing, 15–42
    in frequency domain, 69–73
    intervention analysis, 81–83
    multivariate time series, 78–79
    by PSDA, 43–58
    purpose of estimation, 85
    scale invariance, 58
    sensitivity to initial conditions, 78
    in time series, 41, 59–69, 85
Fractal variance, 59–69, 74
    DFA, 60–63
    Higuchi's fractal dimension, 60,
        65–69
    R/S analysis, 60, 63, 65
    See also Fractality (fractals)
Fractional Brownian motion (fBm),
    55–57
    in DFA, 61, 63

Fractional differencing, ix, 12–13,
    15–42, 58, 86
    ARIMA process, 21, 44
    ARMA (p, q) model, 16
    autocorrelation, removal of, 30–31
    autoregression process, 16
    Brownian motion, 20–21
    challenges, 41–42
    competitive modeling strategies,
        30–36
    detrended data, differencing, 36–38
    differencing parameter, 39–40
    drawbacks, 43, 87
    goodness of fit, 30–31, 33
    Hurst exponent, 40, 88
    integration, 19–21
    long-range dependencies, 22–23,
        24 (figure)
    metastability, 22
    moving average process, 16
    nonstationarity, 21–22, 36
    pink noise, 22
    PSDA advantage over, 55
    random walk, 19–21, 20 (figure)
    to real data, 24–40
    residuals, analysis of, 38,
        39 (figure)
    seasonal patterns, 18–19
    short-range autoregression, 17
        (figure), 18
    spectral regression and, 71–73
    stationarity testing, 21–22, 26–29
    variance reduction, 30
    vs. spectral density analysis, 55
    white noise, 17, 17 (figure)
    See also Power spectral density
        analysis (PSDA)
Fractional Gaussian noise (fGn),
    55–57
    in DFA, 61, 63
Fractions, 41
Frequency domain
    fractality in, 69–73
    time domain to, 44–52
Frequency of cycles, 44
    relative, 44–45, 46 (figure)

Gaussian distribution, 11
Goodness of fit, 30–31, 33
GPH estimator, 70

Harmonic analysis, 13
Higuchi's fractal dimension, 60,
    65–69
Human behavior, 1
Hurst exponent, 40, 67, 69 (table),
    73–74, 88
    antipersistence and, 40
    DFA, 63
    fractal dimension and, 67
    fractional differencing, 40, 88
    periodogram-based regression
        model, 70
    PSDA, 56–57, 57 (table), 88
    R/S analysis, 65
    self-organized criticality and, 90
    spectral regression, 70–73

Interrupted time series, 81–83
Intervention analysis, 81–83
Irregularity, ix, 77, 78

Kwiatkowski–Phillips–Schmidt–Shin
    (KPSS) test, 21–22
    for real data, 26–29

Lag operator, 22
Left–right political orientation.
    See Political orientation in
    Netherlands
Ljung–Box Portmanteau (LBQ) test,
    30–31
Log likelihood (LL) function, 30
Long-range dependencies, 3, 7–9
    fractional differencing, 22–23,
        24 (figure)
    irrregular, 22–23
    regular, 80–81
Lorenz, Edward N., 78

Macropartisanship, 4
Metastability, 22, 40
    periodogram and power spectra,
        50–52, 51 (figure)

time series and ACF plots, 23,
    24 (figure)
Monthly unemployment figures.
    See Unemployment figures in
    United States
Moving average process, 16
Multiplicative time series model, 18
Multivariate time series, 78–79

Nested regularity, 80–81
Nested time series model, 18
Netherlands political orientation.
    See Political orientation in
    Netherlands
Nonstationarity, 21–22, 36, 40, 58
$1/f$ noise, 22

Periodograms, 49, 50–52, 51 (figure)
    in real data, 53–55, 54 (figure)
    regression models, 69–73
    scaled, 48 (figure), 49
Persistence, 22
    Hurst exponent and, 40
Pink noise, 22
    periodogram and power spectra,
        50–52, 51 (figure)
    time series and ACF plots, 23,
        24 (figure)
Political orientation in Netherlands, 24
    detrended data, best fitting models,
        36–37, 38 (table)
    DFA for, 63, 64 (figure)
    fractal dimension, 67, 68 (figure)
    spectral density analysis, 53–55
    stationarity tests, 26–29
    time series and ACF plots, 25–26,
        25 (figure)
Power exponent, 50
Power law, 11, 43, 61, 65
Power of cycles, 44
Power spectra, 43, 50–52, 51 (figure),
    73, 86–88
Power spectral density, 50–52
    in real data, 52–55
Power spectral density analysis
    (PSDA), ix, 12–13, 43–58, 86–88
    advantages, 55, 69, 87

amplitudes, 44–45, 46 (figure)
disadvantages, 58
features, 43, 44
Fourier transform, 46–49
fractional Gaussian noise and
  Brownian motion, 55–57
Hurst exponent, 56, 88
periodograms, 49, 50–52
power law, 11, 43
power spectra, 43, 50–52, 51
  (figure), 73, 86–88
power spectral density, 50–55
purpose of, 43, 49
relative frequencies, 44–45, 46
  (figure), 48
scaled periodogram, 48 (figure), 49
spectral density, 50, 52–55
from time domain to frequency
  domain, 44–52
See also Fractional differencing

Random variability, 17, 17 (figure)
periodogram and power spectra,
  50–52, 51 (figure)
Random walk, 19–21, 20 (figure)
Regular long-range dependencies,
  80–81
Relative frequencies, 44–46, 46
  (figure), 48
Rescaled range (R/S) analysis, 40, 60,
  63, 65, 67, 69 (table)
Residual time series, 38, 39 (figure)
Residual variance, 59–60
R/S analysis. See Rescaled range
  (R/S) analysis

Sampling, 4
Scaled periodogram, 48 (figure), 49
Scale invariance, 59, 60, 61, 86–87
Seasonal patterns, time series analysis,
  18–19
Self-affinity, 66
Self-organized criticality, 90
Self-similarity, 9–11, 10 (figure), 39,
  66, 85, 89–90
Sensitivity to initial conditions, 78

Short-range autoregression, 88
periodogram and power spectra,
  50–52, 51 (figure)
time series and ACF plots, 17
  (figure), 18
Short-range dependencies, 7
Social networks, 12
Spectral density. See Power spectral
  density
Spectral density function, 69
Spectral regression, 69–73, 89
Stationarity, 19, 21–22, 58
Stationary increments, 21
Statistical dependency, 15
Statistical equilibrium, 19

TBATS model. See Trigonometric
  Box–Cox ARMA Trend Seasonal
  (TBATS) model
Teen pregnancies in Texas, 7, 8
  (figure), 24
comparative modeling, 36,
  36 (table)
fractional differencing, 29,
  29 (table)
periodogram and power spectra, 53,
  54 (figure)
residual time series, 38, 39 (figure)
short- and long-range nonrandom
  patterns, 80–81, 80 (figure)
spectral density analysis, 53–55
stationarity tests, 26–29
time series and ACF plots, 25
  (figure), 26, 80–81, 80 (figure)
Tension-release pattern, 11
Time series analysis, 2
ARIMA, 21
ARMA (p, q) model, 16
autocorrelation in, 2–3
autoregression process, 16
complexity, 9–12
dependency between observations, 4
fractality in, 59–69
fractional differencing, 15–42
to frequency domain, 44–52
interrupted time series, 81–83

long-range dependencies, 3, 7–9
moving average process, 16
multiplicative/nested, 18
multivariate, 78–79
nonrandomness in, 4–7
nonstationarity, 21–22
pink noise, 22, 23, 24 (figure)
PSDA, 43–58
purposes of, 15
random variability, 17, 17 (figure)
random walk, 19–21, 20 (figure)
residual, 38, 39 (figure)
results in, 15–22
seasonal patterns, 18–19
short-range autoregression, 17
    (figure), 18
short-range dependencies, 7
stationarity, 19, 21–22
traditional, 2–3, 4–7
United States political system
    stability, 4–7, 82–83
variance components, 16
white noise, 17, 17 (figure)
Traditional time series analysis, 2–3, 4–7
Trigonometric Box–Cox ARMA Trend
    Seasonal (TBATS) model, 80

Unemployment figures in United
    States, 24
comparative modeling, 31–33,
    32 (table)
DFA, 63, 64 (figure)
periodogram and power spectra, 53,
    54 (figure)
spectral density analysis, 53–55
stationarity tests, 26–29
time series and ACF plots, 25
    (figure), 26
United States
monthly unemployment figures in.
    See Unemployment figures in
    United States
presidential approval, consumer
    sentiment, and
    macropartisanship in, 4–7, 5
    (figure), 8, 82–83
teen pregnancies in Texas. See Teen
    pregnancies in Texas
Univariate time series, 77

White noise, 17, 17 (figure)
periodogram and power spectra,
    50–52, 51 (figure)